NATURAL DISASTERS:
Protecting the Public's Health

Scientific Publication No. 575

Pan American Health Organization
Pan American Sanitary Bureau, Regional Office of the
World Health Organization
525 Twenty-third Street, N.W.
Washington, D.C. 20037, USA

Also published in Spanish with the title:
Los desastres naturales y la protección de la salud
ISBN 92 75 31575 2

PAHO Library Cataloguing in Publication Data

Pan American Health Organization
Natural disasters: Protecting the public's health.
Washington, D.C. : PAHO, ©2000.
xi, 119 p.—(Scientific Publication, 575)

ISBN 92 75 11575 3

I. Title   II. (Series)
1. NATURAL DISASTERS 2. HEALTH EFFECT OF DISASTERS
3. DISASTERS PLANNING — organization and administration
4. EMERGENCIES IN DISASTERS — organization and administration
5. INTERNATIONAL COOPERATION

NLM HV553

3/01

CONTENTS

PREFACE

Two decades have passed since the Pan American Health Organization published the first edition of these guidelines. In the intervening years, disaster prevention, mitigation, and preparedness has evolved in important ways. Clearly, it was time for us to revisit this publication.

Twenty years ago, disaster management was simply left to a few dedicated professionals. Roles were clear: rescue workers rushed to help victims and certain agencies stepped in to provide temporary shelter and food. And society at large, a while after the impact, erased the disaster from its memory—until the next one came to wreak new destruction.

Unfortunately, disasters in the Americas and throughout the world have provided ample opportunities to test the policies and recommendations set out nearly twenty years ago. Over time, the approach has changed.

Today, the management of humanitarian assistance involves many more and different players, and disasters are recognized as public health priorities in which the health system plays a significant role. Today, prevention, mitigation, and preparedness are part of the vocabulary of disaster administrators in national and international organizations and, more importantly, they are used to advance the cause of disaster reduction. Today, society's involvement in disasters both precedes the impact and remains alive long after. Finally, the interrelationship between human development and disasters is better understood today—how disasters can permanently damage a country's economy, but, at the same time, how the path toward development may put a country at greater risk to the destructive consequences of natural disasters.

We are pleased to offer these updated guidelines. They include all the principles and recommendations that have withstood the test of time and new concepts and understanding gleaned along the way. May they strengthen disaster prevention, mitigation, and preparedness in our countries. May they save lives.

George A.O. Alleyne
Director

INTRODUCTION

This publication echoes the 1981 *Emergency Health Management after Natural Disaster* (Scientific Publication No. 407), but it is a whole new book, with a fresh organization and much new content. It sketches the role the health sector plays in reducing the impact of disasters and describes how it can carry it out.

These guidelines aim at presenting a framework that an administratror can rely on to make effective decisions in managing the health sector's activities to reduce the consequences of disasters. It does not pretend to cover every contingency. In fact, some of the suggested procedures may need to be adapted to fill some local needs. We hope that this book will help to develop manuals that can be tailored to local conditions.

The book's 14 chapters and 4 technical annexes describe the general effects of disasters on health, highlighting myths and realities. Although every disaster is unique, there are common features that can be used to improve the management of humanitarian assistance in health and the use of available resources.

Chapter 2 is one of the main innovations. It summarizes how the health sector must structure itself and work with other sectors to cope with disasters. The chapter covers the health sector's activities for reducing the consequences of disasters that affect response, preparedness, and mitigation phases, pointing out where these are interdependent.

Chapter 3 deals with disaster preparedness—its multisectoral nature and its specific application in the health sector. It sets forth guidelines for preparing health sector plans, means of coordination, and special technical programs that cover every aspect of normal operations before a disaster hits.

Chapter 4 also includes new material. It deals with the disaster mitigation activities that the health sector must promote and put in place. Mitigation measures are designed to reduce the vulnerability to disasters in health establishements (including drinking water and sewerage systems) and to reduce the magnitude of the disaster's effects. Mitigation activities complement preparedness and response activities.

Chapter 5 deals with the response to disasters, as well as its coordination and the evaluation of health needs. Chapters 6 through 11, and Chapter 14, retain the organization of the 1981 guidelines, but they have been updated. Chapter 12, dealing with humanitarian supplies, and Chapter 13, dealing with humanitarian assistance also have been revised in-depth.

Finally, two of the four annexes—the one dealing with the management of supplies and the one dealing with the national mitigation program—are entirely new; the remaining two have been updated.

This book is aimed primarily at health sector professionals who participate in disaster preparedness, response, and mitigation. The intersectoral perspective is now so essential, however, that anyone interested in disaster reduction will find here a useful primer. Public health students and professors also can rely on this book as a manual for formal or informal courses.

ACKNOWLEDGMENTS

The drafting and technical review of this publication has benefited from the co-operation of many persons, both within and outside the Pan American Health Organization.

This book would not have been possible without the excellent work of the following experts at the Pan American Health Organization: Dr. Claude de Ville de Goyet, Dr. Hugo Prado Monje, Dr. Jean Luc Poncelet, Dr. Luis Jorge Pérez, Dr. Dana Van Alphen, Dr. José Luis Zeballos, Mr. Claudio Osorio, Mr. Adrianus Vlugman, Mrs. Patricia Bittner, and Mr. Ricardo Pérez. Dr. Miguel Gueri, expert in food and nutrition; Dr. Edgardo Acosta, executive director of FUNDESUMA; and Mr. John Scott, expert in telecommunications and disasters, also made invaluable contributions.

We also wish to thank many other experts for their important comments and critiques: Dr. Stephen J. Corber, Dr. Gustavo Bergonzoli, Dr. Alejandro Santander, Dra. Rocio Saenz, and Mr. Homero Silva, all from PAHO/WHO, and Dr. Raul Morales Soto, from Peru, and Mr. Edgardo Quiros, consultant at FUNDESUMA.

The excellent contributions contributions from other WHO regions and other international organizations have given this book global scope. We are especially grateful fo Dr. David L. Heyman, Dr. M.C. Thuriaux, Dr. Maria Neira, Dr; Eric K. Noji, and Dr. Xavier Leus, all from WHO.

Many others from national, subregional, and regional organizations, mostly from the Region of the Americas enriched this text with their comments and recommendations. To all of them, many, many thanks.

Finally, we wish to acknowledge Liz Stonaker's untiring and diligent editorial work.

CHAPTER 1.
GENERAL EFFECTS OF DISASTERS ON HEALTH

In the past, sudden-impact disasters were believed to cause not only widespread death, but also massive social disruption and outbreaks of epidemic disease and famine, leaving survivors entirely dependent on outside relief. Systematic observation of the effects of natural disasters on human health has led to different conclusions, both about the effects of disaster on health and about the most effective ways of providing humanitarian assistance.

The term "disaster" usually refers to the natural event (e.g., a hurricane or earthquake) in combination with its damaging effects (e.g., the loss of life or destruction of buildings). "Hazard" refers to the natural event, and "vulnerability" to the susceptibility of a population or system (e.g., a hospital, water supply and sewage system, or aspects of infrastructure) to the effects of the hazard. The probability that a particular system or population will be affected by hazards is known as the "risk." Hence, risk is a function of the vulnerability and the hazard, and is expressed as follows:

$$Risk = Vulnerability \times Hazard$$

Though all disasters are unique in that they affect areas with different levels of vulnerability and with distinct social, health, and economic conditions, there are still similarities between disasters. If recognized, these common factors can be used to optimize the management of health humanitarian assistance and use of resources (see Table 1.1). The following points should be noted:

1. There is a relationship between the type of disaster and its effect on health. This is particularly true of the immediate impact in causing injuries. For example, earthquakes cause many injuries requiring medical care, while floods and tidal waves cause relatively few.
2. Some effects are a potential, rather than an inevitable, threat to health. For example, population movement and other environmental changes may lead to increased risk of disease transmission, although epidemics generally do not result from natural disasters.
3. The actual and potential health risks after a disaster do not all occur at the same time. Instead, they tend to arise at different times and to vary in importance within a disaster-affected area. Thus, casualties occur mainly at the time and place of impact and require immediate medical care, while the risks of increased disease transmission take longer to develop and are greatest where there is overcrowding and standards of sanitation have declined.

TABLE 1.1. Short-term effects of major disasters.

Effect	Earthquakes	High winds (without flooding)	Tidal waves/flash floods	Slow-onset floods	Landslides	Volcanoes/Lahars
Deaths[a]	Many	Few	Many	Few	Many	Many
Severe injuries requiring extensive treatment	Many	Moderate	Few	Few	Few	Few
Increased risk of communicable diseases	Potential risk following all major disasters (Probability rising with overcrowding and deteriorating sanitation)					
Damage to health facilities	Severe (structure and equipment)	Severe	Severe but localized	Severe (equipment only)	Severe but localized	Severe (structure and equipment)
Damage to water systems	Severe	Light	Severe	Light	Severe but localized	Severe
Food shortage	Rare (may occur due to economic and logistic factors)		Common	Common	Rare	Rare
Major population movements	Rare (may occur in heavily damaged urban areas)		Common (generally limited)			

[a]Potential lethal impact in absence of preventive measures.

4. Disaster-created needs for food, shelter, and primary health care are usually not total. Even displaced people often salvage some of the basic necessities of life. Furthermore, people generally recover quickly from their immediate shock and spontaneously engage in search and rescue, transport of the injured, and other private relief activities.
5. Civil wars and conflicts generate a distinct set of public health problems and operational constraints. They are not covered in any depth in this publication.

Effective management of health humanitarian aid depends on anticipating and identifying problems as they arise, and delivering specific materials at the precise times and points where they are needed. The logistical ability to transport maximum numbers of supplies and personnel from abroad to disaster areas in Latin America and the Caribbean is less essential. Cash is the most effective donation, particularly since it can be used to purchase supplies locally.

HEALTH PROBLEMS COMMON TO ALL NATURAL DISASTERS

Social Reactions

After a major natural disaster, behavior only rarely reaches generalized panic or stunned waiting. Spontaneous yet highly organized individual action accrues as survivors rapidly recover from their initial shock and set about purposefully to achieve clear personal ends. Earthquake survivors often begin search and rescue activities minutes after an impact and within hours may have organized themselves into groups to transport the injured to medical posts. Actively antisocial behavior such as widespread looting occurs only in exceptional circumstances.

Although everyone thinks his or her spontaneous reactions are entirely rational, they may be detrimental to the community's higher interests. A person's conflicting roles as family head and health official, for instance, have in some instances resulted in key relief personnel not reporting to duty until their relatives and property are safe.

Rumors abound, particularly of epidemics. As a result, considerable pressure may be put on the authorities to undertake emergency humanitarian work such as mass vaccinations against typhoid or cholera, without sound medical justification. In addition, people may be reluctant to submit to measures that the authorities think necessary. During warning periods, or after the occurrence of natural disasters, people are reluctant to evacuate, even if their homes are likely to be or have been destroyed.

These patterns of behavior have two major implications for those making decisions about humanitarian programs. First, patterns of behavior and demands for emergency assistance can be limited and modified by keeping the population informed and by obtaining necessary information before embarking on extended relief programs. Second, the population itself will provide most rescue and first aid, take the injured to hospitals if they are accessible, build temporary shelters, and carry out other essential tasks. Additional resources should, therefore, be directed toward meeting the needs that survivors themselves cannot meet on their own.

Communicable Diseases

Natural disasters do not usually result in massive outbreaks of infectious disease, although in certain circumstances they do increase the potential for disease transmission. In the short-term, the most frequently observed increases in disease incidence are caused by fecal contamination of water and food; hence, such diseases are mainly enteric.

The risk of epidemic outbreaks of communicable diseases is proportional to population density and displacement. These conditions increase the pressure on water and food supplies and the risk of contamination (as in refugee camps), the disruption of preexisting sanitation services such as piped water and sewage, and the failure to maintain or restore normal public health programs in the immediate post-disaster period.

In the longer term, an increase in vector-borne diseases occurs in some areas because of disruption of vector control efforts, particularly following heavy rains and floods. Residual insecticides may be washed away from buildings and the number of mosquito breeding sites may increase. Moreover, displacement of wild or domesticated animals near human settlements brings additional risk of zoonotic infections.

In complex disasters where malnutrition, overcrowding, and lack of the most basic sanitation are common, catastrophic outbreaks of gastroenteritis (caused by cholera or other diseases) have occurred, as in Rwanda/Zaire in 1994.

Population Displacements

When large, spontaneous or organized population movements occur, an urgent need to provide humanitarian assistance is created. People may move to urban areas where public services cannot cope, and the result may be an increase in morbidity and mortality. If much of the housing has been destroyed, large population movements may occur within urban areas as people seek shelter with relatives and friends. Surveys of settlements and towns around Managua, Nicaragua, following the December 1972 earthquake indicated that 80% to 90% of the 200,000 displaced persons were living with relatives and friends; 5% to 10% were living in parks, city squares, and vacant lots; and the remainder were living in schools and other public buildings. Following the earthquake that struck Mexico City in September 1985, 72% of the 33,000 homeless found shelter in areas close to their destroyed dwellings.

In internal conflicts, such as occurred in Central America (1980s) or Colombia (1990s), refugees and internally displaced populations are likely to persist.

Climatic Exposure

The health hazards of exposure to the elements are not great, even after disasters in temperate climates. As long as the population is dry, reasonably well clothed, and able to find windbreaks, death from exposure does not appear to be a major risk in Latin America and the Caribbean. The need to provide emergency shelter therefore varies greatly with local conditions.

Food and Nutrition

Food shortages in the immediate aftermath may arise in two ways. Food stock destruction within the disaster area may reduce the absolute amount of food available, or disruption of distribution systems may curtail access to food, even if there is no absolute shortage. Generalized food shortages severe enough to cause nutritional problems do not occur after earthquakes.

Flooding and sea surges often damage household food stocks and crops, disrupt distribution, and cause major local shortages. Food distribution, at least in the short term, is often a major and urgent need, but large-scale importation/donation of food is not usually necessary.

In extended droughts, such as those occurring in Africa, or in complex disasters, the homeless and refugees may be completely dependent on outside sources for food supplies for varying periods of time. Depending on the nutritional condition of these populations, especially of more vulnerable groups such as pregnant or lactating women, children, and the elderly, it may be necessary to institute emergency feeding programs.

Water Supply and Sanitation

Drinking water supply and sewerage systems are particularly vulnerable to natural hazards, and the disruptions that occur in them pose a serious health risk. The systems are extensive, often in disrepair, and are exposed to a variety of hazards. Deficiencies in established amounts and quality of potable water and difficulties in the disposal of excreta and other wastes result in the deterioration of sanitation, contributing to conditions favorable to the spread of enteric and other diseases.

Mental Health

Anxiety, neuroses, and depression are not major, acute public health problems immediately following disasters, and family and neighbors in rural or traditional societies can deal with them temporarily. A group at high risk, however, seems to be the humanitarian volunteers or workers themselves. Wherever possible, efforts should be made to preserve family and community social structures. The indiscriminate use of sedatives and tranquilizers during the emergency relief phase is strongly discouraged. In industrialized or metropolitan areas in developing countries, mental health problems are reported to be significant during long-term rehabilitation and reconstruction and need to be dealt with during that phase.

Damage to the Health Infrastructure

Natural disasters can cause serious damage to health facilities and water supply and sewage systems, having a direct impact on the health of the population dependent on these services. In the case of structurally unsafe hospitals and health centers, natural disasters jeopardize the lives of occupants of the buildings, and limit the capacity to provide health services to disaster victims. The earthquake that struck Mexico City in 1985 resulted in the collapse of 13 hospitals. In just three of those buildings, 866 people died, 100 of whom were health personnel. Nearly 6,000 hospital beds were lost in the metropolitan facilities. As a result of Hurricane Mitch in 1998, the water supply systems of 23 hospitals in Honduras were damaged or destroyed, and 123 health centers were affected. Peru reported that nearly 10% of the country's health facilities suffered damage as a result of El Niño events in 1997–1998.

IMMEDIATE HEALTH PROBLEMS RELATED TO THE TYPE OF DISASTER

Earthquakes

Usually because of dwelling destruction, earthquakes may cause many deaths and injure large numbers of people. The toll depends mostly on three factors.

The first factor is housing type. Houses built of adobe, dry stone, or unreinforced masonry, even if only a single story high, are highly unstable and their collapse causes many deaths and injuries. Lighter forms of construction, especially woodframe, have proved much less dangerous. After the 1976 earthquake in Guatemala, for example, a survey showed that in one village with a population of 1,577, all of those killed (78) and severely injured had been in adobe buildings, whereas all residents of woodframe buildings survived. In the earthquake affecting the villages of Aiquile and Totora in Bolivia in 1998, 90% of deaths resulted from the collapse of adobe housing.

The second factor is the time of day at which the earthquake occurs. Night occurrence was particularly lethal in the earthquakes in Guatemala (1976) and Bolivia (1998), where most damage occurred in adobe houses. In urban areas with well-constructed housing but weak school or office structures, earthquakes occurring during the day result in higher death rates. This was the case in the 1997 earthquake that struck the towns of Cumaná and Cariaco, Venezuela. In Cumaná an office building collapsed, and in Cariaco two schools collapsed, accounting for most of the dead and injured.

The last factor is population density: the total number of deaths and injuries is likely to be much higher in densely populated areas.

There are large variations within disaster-affected areas. Mortality of up to 85% occasionally may occur close to the epicenter of the earthquake. The ratio of dead to injured decreases as the distance from the epicenter increases.

Some age groups are more affected than others; fit adults are spared more than small children and the elderly, who are less able to protect themselves. However, 72% of the deaths resulting from collapsed buildings in the 1985 Mexico earthquake were among persons between the ages of 15 and 64 (see Table 1.2).

Secondary disasters may occur after earthquakes and increase the number of casualties requiring medical attention. Historically, the greatest risk is from fire, although in recent decades, post-earthquake fires causing mass casualties have been uncommon. However, in the aftermath of the earthquake that hit Kobe, Japan, in 1995, over 150 fires occurred. Some 500 deaths were attributed to fires, and approximately 6,900 structures were damaged. Fire-fighting efforts were hindered because streets were blocked by collapsed buildings and debris, and the water system was severely damaged.

Little information is available about the kinds of injuries resulting from earthquakes, but regardless of the number of casualties, the broad pattern of injury is likely to be a mass of injured with minor cuts and bruises, a smaller group suffering from simple fractures, and a minority with serious multiple fractures or internal injuries requiring surgery and other intensive treatment. For example, after the 1985 earthquake in Mexico, 1,879 (14.9%) of the 12,605 patients treated by the emergency medical services (including certain routine cases) needed hospitalization, most of them for a 24-hour period.

Most of the demand for health services occurs within the first 24 hours of an event. Injured persons may continue to show up at medical facilities only during the first three to five days, after which presentation patterns return almost to nor-

TABLE 1.2. Distribution of deaths by age group resulting from the September 1985 earthquake in Mexico City.[a]

Age group (years)	Deaths	Percent of deaths
Under 1 year old	173	4.8
1 – 4	143	4.0
5 – 14	287	8.0
15 – 24	770	21.5
25 – 44	1,293	36.1
45 – 64	519	14.5
65 and older	226	6.3
Not defined	168	4.7
Total	3,579	100

[a]Bodies recovered from collapsed buildings between 19 September and 29 October 1985.
Source: General Directorate of Investigation, Attorney General, Department of Justice, Mexico, D.F.

mal. A good example of the crucial importance of the timing of emergency care is seen in the number of admissions to a field hospital after the 1976 earthquake in Guatemala, as shown in Figure 1.1. From day six onward, admissions fell dramatically, despite intensive case-finding in remote rural areas.

Patients may appear in two waves, the first consisting of casualties from the immediate area around the medical facility and the second of referrals as humanitarian operations in more distant areas become organized.

Destructive Winds

Unless they are complicated by secondary disasters such as the floods or sea surges often associated with them, destructive winds cause relatively few deaths and injuries. Effective warning before such windstorms will limit morbidity and mortality, and most injuries will be relatively minor. Most of the public health consequences from hurricanes and tropical storms result from torrential rains and floods, rather than wind damage. The catastrophic death toll—an estimated 10,000—in Central American countries after Hurricane Mitch in 1998 was primarily caused by flooding and mudslides.

Flash Floods, Sea Surges, and Tsunamis

These phenomena may cause many deaths, but leave relatively few severely injured in their wake. Deaths result mainly from drowning and are most common among the weakest members of the population. More than 50% of the deaths in

FIGURE 1.1. Admissions and occupancy rates at the field hospital in Chimaltenango, Guatemala, 1976.

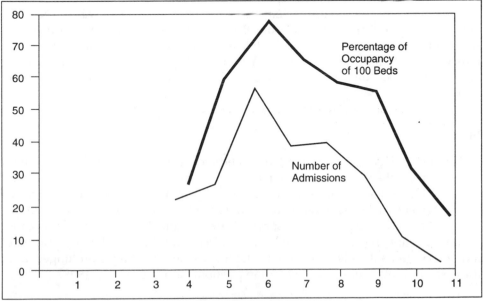

Nicaragua following Hurricane Mitch in 1998 were due to flash floods and mudslides on the slopes of the Casitas Volcano.

Volcanoes

Volcanoes are found worldwide and significant numbers of people often live in close proximity to them. The fertile volcanic soil is good for agriculture and is attractive for the establishment of towns and villages. In addition, volcanoes have long periods of inactivity, and some generations have no experience with volcanic eruptions, thereby encouraging the population to feel some degree of security in spite of the danger in living close to a volcano. The difficulty in predicting a volcanic eruption compounds the situation.

Volcanic eruptions affect the population and infrastructure in many ways. Immediate trauma injuries may be caused if there is contact with volcanic material. The super-heated ash, gases, rocks, and magma can cause burns severe enough to kill immediately. Falling rocks and boulders also can result in broken bones and other crush-type injuries. Breathing the gases and fumes can cause respiratory distress.

Health facilities and other infrastructure can be destroyed in minutes if they lie in the path of pyroclastic flows and lahars (mudflows containing volcanic debris). Accumulated ash on roofs can greatly increase the risk of collapse. Contamination of the environment (e.g., water and food) with volcanic ash also can disrupt environmental health conditions; this effect is compounded when the population must be evacuated and housed in temporary shelters.

If the eruptive phase is prolonged, as in the case on the Caribbean island of Montserrat where the Soufriere Hills volcano began erupting in July 1995 and continued for several years, other health effects, such as increased stress and anxiety in the remaining population, become important. Long-term inhalation of silica-rich ash also can result in pulmonary silicosis years later.

One of the most devastating events to occur in Latin America was the November 1985 eruption of the Nevado del Ruiz volcano in Colombia. The heat and seismic forces melted a portion of the icecap on the volcano, resulting in an extensive lahar that buried the city of Armero, killing 23,000 people and injuring 1,224. Some 1,000 km^2 of prime agricultural land at the base of the volcano were affected.

Floods

Slow-onset flooding causes limited immediate morbidity and mortality. A slight increase in deaths from venomous snakebites has been reported, but not fully substantiated. Traumatic injuries caused by flooding are few and require only limited medical attention. While flooding may not result in an increased frequency of disease, it does have the potential to spark communicable disease outbreaks because of the interruption of basic public health services and the overall deterioration of living conditions. This is of concern particularly when flooding is prolonged, as in the case of events caused by the El Niño phenomenon in 1997 and 1998.

Landslides

Landslides have become an increasingly common disaster in Latin America and the Caribbean; intense deforestation, soil erosion, and construction of human settlements in landslide-prone areas have resulted in catastrophic events in recent years. This has been the case in both urban and rural areas. Rain brought by Tropical Storm Bret triggered landslides in poor neighborhoods on the outskirts of Caracas, Venezuela, in August 1993. At least 100 people died, and 5,000 were left homeless. High death tolls occurred in the gold mining town of Llipi, Bolivia, in 1992, where a landslide buried the entire village, killing 49. Deforestation contributed significantly to the disaster, and mining tunnels collapsed. A similar disaster occurred in the gold mining region of Nambija, Ecuador, in 1993, claiming 140 lives.

In general, this phenomenon causes high mortality, although injuries are few. If there are health structures (hospitals, health centers, water systems) in the path of the landslide, they can be severely damaged or destroyed.

MYTHS AND REALITIES OF NATURAL DISASTERS

Many mistaken assumptions are associated with the impact of disasters on public health. Disaster planners and managers should be familiar with the following myths and realities:

Myth: Foreign medical volunteers with any kind of medical background are needed.

Reality: The local population almost always covers immediate lifesaving needs. Only medical personnel with skills that are not available in the affected country may be needed.

Myth: Any kind of international assistance is needed, and it's needed immediately!

Reality: A hasty response that is not based on an impartial evaluation only contributes to the chaos. It is better to wait until genuine needs have been assessed. In fact, most needs are met by the victims themselves and their local government and agencies, not by foreign intervenors.

Myth: Epidemics and plagues are inevitable after every disaster.

Reality: Epidemics do not spontaneously occur after a disaster and dead bodies will not lead to catastrophic outbreaks of exotic diseases. The key to preventing disease is to improve sanitary conditions and educate the public.

Myth: Disasters bring out the worst in human behavior (e.g., looting, rioting).

Reality: Although isolated cases of antisocial behavior exist, most people respond spontaneously and generously.

Myth: The affected population is too shocked and helpless to take responsibility for their own survival.

Reality: On the contrary, many find new strength during an emergency, as evidenced by the thousands of volunteers who spontaneously united to sift through the rubble in search of victims after the 1985 Mexico City earthquake.

Myth: Disasters are random killers.

Reality: Disasters strike hardest at the most vulnerable groups—the poor, especially women, children, and the elderly.

Myth: Locating disaster victims in temporary settlements is the best alternative.

Reality: It should be the last alternative. Many agencies use funds normally spent for tents to purchase building materials, tools, and other construction-related support in the affected country.

Myth: Things are back to normal within a few weeks.

Reality: The effects of a disaster last a long time. Disaster-affected countries deplete much of their financial and material resources in the immediate post-impact phase. Successful relief programs gear their operations to the fact that international interest wanes as needs and shortages become more pressing.

Chapter 2.
Structuring Health Disaster
Management

The role of disaster professionals in Latin America and the Caribbean has changed considerably over the last 30 years. Up to the 1970s, their actions were mostly limited to the disaster aftermath, or disaster response. However, the ministries of health and other governmental and nongovernmental organizations of the Region, recognizing that a number of relief operations were poorly coordinated, started to work on disaster preparedness to provide better humanitarian assistance to their populations.

Following the devastation caused by the 1985 earthquake in Mexico City, and with particular concern for the losses suffered in hospitals, regional authorities acknowledged that not only did the population need assistance in the disaster aftermath, but they deserved to have a less vulnerable health system. With existing technology it is possible, at a reasonable cost, to greatly reduce the susceptibility of a system to the effects of a hazard. This approach was strongly reinforced by the U.N. General Assembly's designation of the 1990s as the "International Decade of Natural Disaster Reduction," and spurred concerted efforts in the Region to implement disaster mitigation programs.

There are three fundamental aspects of disaster management:

- disaster response,
- disaster preparedness, and
- disaster mitigation.

These three aspects of disaster management correspond to phases in the so-called "disaster cycle" (see Figure 2.1).

Activities in the aftermath of disaster include response, rehabilitation, and reconstruction. Chapter 5 of this book concentrates on the coordination of health activities in the emergency period and some aspects of rehabilitation. There will be few references to reconstruction. During the reconstruction phase, the coordination mechanism, project approval, and other decisions are taken in an environment that is much closer to the "normal" situation. Time is no longer the most important factor. The reconstruction period provides an opportunity to implement the health sector's disaster mitigation programs and to initiate or reinforce disaster preparedness programs (see Chapters 3 and 4).

FIGURE 2.1. Management sequence of a sudden-onset disaster.

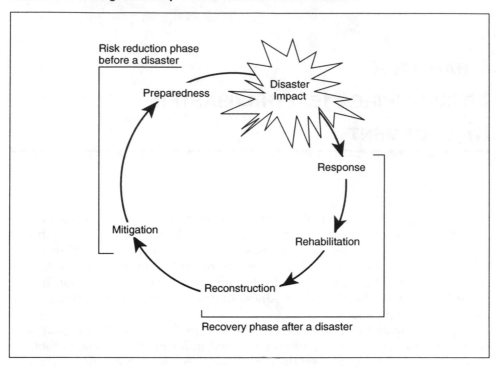

NATIONAL DISASTER MANAGEMENT AGENCIES

Many Latin American and Caribbean countries have developed national disaster management agencies to coordinate activities involved in disaster preparedness, mitigation, response, and rehabilitation. These organizations are usually assigned to the ministries of defense or interior, or their national equivalent.

Disaster preparedness and mitigation have few immediately visible benefits, and until a major catastrophe occurs, the results are hidden. The establishment of national disaster management agencies is a complex and slow process, requiring consistent political and public support. Legislation is required to ensure the continuity of these agencies and to guarantee their funding. Despite these difficulties, institutionalized disaster management programs are the only long-term solutions to reducing the vulnerability of all sectors of society to disasters.

While the health sector can do a great deal on its own to improve its ability to respond in a disaster situation, it is impossible to reduce the impact of hazards without the involvement of public works, financing, education, housing, and other sectors. The health sector should take the lead in promoting the institutionalization of a national disaster management agency and will benefit from decisions taken to reduce vulnerability in other sectors. For example, a strictly enforced building code will reduce the numbers of homes and other structures damaged by an earthquake, and consequently reduce the number of casualties needing treatment.

NATIONAL HEALTH DISASTER MANAGEMENT PROGRAM

The health sector in Latin America and the Caribbean has been working to reduce its vulnerability to disasters by establishing national health disaster management programs. Responsibilities of such programs encompass the entire health sector, and not only the ministry of health. The program must play the leading role in promoting and coordinating prevention, mitigation, preparedness, response, and early rehabilitation activities related to health. The scope of the program is a multi-hazard one, including all large-scale emergencies regardless of their origin (natural disasters, chemical or radiation accidents, civil conflicts, violence, etc.).

As part of preparedness activities, the health disaster program should ensure that disaster plans are in place and up-to-date (see Chapter 3). To test these plans, simulation exercises involving the health and other sectors must be conducted on a regular basis.

Activities related to mitigating the effects of disasters (see Chapter 4) require the inclusion of mitigation measures in all ministry of health programs (development of new services, construction or remodeling of facilities, and maintenance programs for hospitals and other health facilities). Vulnerability reduction also must be promoted for services that affect health, especially in water supply and sewerage systems. Finally, mitigation activities require the development of construction standards and norms to ensure that health facilities will resist potential hazards.

Finally, during the disaster response phase, the health disaster management program coordinates all health sector activities to reduce the loss of life and property and protect the integrity of health services (see Chapters 5 through 13).

Regional and Local Responsibilities

This chapter focuses on responsibilities of the health disaster management program at the national or central level, but the same principles apply to the heads of disaster programs at provincial or regional levels as well as in epidemiology and environmental health departments, hospitals, social security institutions, water services, and NGOs.

HEALTH DISASTER COORDINATOR

Establishing a single focal point for coordination of disaster-related health activities ensures the optimal use of health resources available to the ministry of health, social security agency, armed forces, and the private sector. A full-time Health Disaster Coordinator should be designated, and, as in the case of most Latin American and Caribbean countries, assigned to the highest level of the ministry of health (e.g., in the advisory unit of the Minister or Vice-Minister), or to the Director-General of Health, especially when that division has operational responsibilities for emergency health care.

The Health Disaster Coordinator is responsible for the development of the health sector's disaster preparedness, mitigation, and prevention programs. The characteristics of his/her duties have evolved in recent years in Latin America and the Caribbean from those of a low-profile troubleshooter to a multifaceted skilled professional responsible for leading the national health disaster management program.

As can be seen from the functions described in Box 2.1, the responsibilities of the national health disaster management program are complex. Most programs now have one person specialized in response and preparedness and another assigned to disaster mitigation.

Coordination of all components—public and private—of the health sector requires that a standing, advisory committee be established. Coordinated by the Health Disaster Coordinator, the committee includes health sector specialists (epidemiology, environmental health, hospital administration) and representatives of major government agencies providing health services, the Red Cross, NGOs, as well as representatives of the international community involved in health activities. The committee meets on a regular basis to take operational decisions and to coordinate their agencies' activities regarding humanitarian assistance.

ASSESSING RISK

To appropriately orient the activities of the disaster management program, the Health Disaster Coordinator must have an understanding of the risks (hazard and vulnerability) present in the country under his or her responsibility. Hazard probability and vulnerability of systems change constantly, depending on scientific information and development processes of the country (for example, growth of cities, changes in building codes, and installation of new industries). The activities of the preparedness and mitigation subprograms are heavily dependent on risk assessment. The main elements to be taken into account for those subprograms are described in Chapters 3 and 4, respectively.

The identification of risks posed by natural hazards and those caused by human activity requires collaboration of the health sector with the scientific community (seismologists, meteorologists, social scientists); environmental specialists; engineers; urban planners; fire brigades; private industry; and in the case of complex emergencies, political entities.

There are certain rules for estimating risk (see Chapter 4), but the level of risk that is deemed tolerable is less precise and is dependent on factors such as cultural and social patterns, public and political awareness, and financial constraints. The Health Disaster Coordinator must understand what is considered acceptable risk to determine priorities in the disaster management program.

TRAINING

Training in all components of the disaster management program is necessary if activities are to be properly implemented. The failures in disaster mitigation, preparedness, and response are largely due to the gaps that exist between different professions and a lack of specific training for health care and public health personnel. Many health professionals have never received training, experienced a disaster situation, or participated in disaster management activities. Professionals employed in other sectors such as public works, financing (involved in construction of health facilities), foreign affairs, or the national disaster management agency (humanitarian assistance) should be aware of disaster preparedness and mitigation issues as they relate to the health sector.

BOX 2.1. The National Health Disaster Management Program.

The program's areas of responsibility are promotion, establishment of standards, training, and coordination with other institutions and sectors, as outlined below.

Promotion
- Health and social aspects and benefits of disaster management with other sectors, including the private sector;
- Inclusion of disaster reduction into development activities of other programs and divisions of the ministry of health and other health sector institutions; and
- Public education through mass media and health educators.

Establishment of standards
- Building and maintenance standards for health facilities in disaster prone areas, taking into consideration mitigation and preparedness measures;
- Norms for contingency planning, simulation exercises, and other preparedness activities in the health sector;
- Lists of essential drugs and supplies for emergencies; and
- Standardized telecommunication protocols.

Training
- In-service training of health personnel (from disaster prevention to response);
- Promotion of disaster management in the curricula of undergraduate and graduate schools in health sciences (such as schools of medicine, nursing, and environmental health); and
- Inclusion of health related topics in disaster management training for other sectors (e.g., planning and foreign affairs).

Collaboration with other institutions and sectors
- The national disaster management agency or other agency with multisectoral responsibility;
- Disaster focal point or commission in other sectors (e.g., national disaster management agency, legislature, foreign affairs division, public works departments, NGOs);
- Disaster programs in the health sector in and outside the country, particularly in neighboring countries or territories; and
- Relief organizations at the national or international level (bilateral and UN agencies, NGOs).

In the event of disaster, the program is responsible for:
- Mobilization of the health response; and
- Providing advice and coordinating operations on behalf of the head of the health sector (minister of health), and supporting the health response in case of large-scale emergencies resulting from natural, technological, or man-made disasters.

The health disaster management program is responsible for promoting ongoing training in health disaster management. The two main approaches to accomplishing this are: (a) continuous training at the institutional level, and (b) academic training provided by a large variety of institutions at the undergraduate and graduate levels or through continuing education.

EVALUATION OF THE HEALTH DISASTER MANAGEMENT PROGRAM

Health disaster management programs are evaluated based on the objectives and functions that have been specifically assigned to it.[1] However, the three indicators that follow are useful for evaluation and can be adapted.

Evaluation of the Preparedness Program

- Date of last review of the national and regional disaster plans of the ministry of health and water system authority;
- Annual disaster exercise, test of disaster plan.

Evaluation of Mitigation Measures

- Percentage of health facilities and water supply and sewerage system having undergone vulnerability study;
- Percentage of critical health facilities and water supply and sewerage system that will remain functional after disasters.

Evaluation of the Training Program

- Number of persons with specialized disaster management training;
- Number of hours of disaster management training available at the undergraduate and postgraduate levels.

---

[1]See Pan American Health Organization, Emergency Preparedness Program, *Guidelines for Assessing Disaster Preparedness in the Health Sector* (Washington, D.C., 1995).

CHAPTER 3.
DISASTER PREPAREDNESS

The objective of disaster preparedness is to ensure that appropriate systems, procedures, and resources are in place to provide prompt, effective assistance to disaster victims, thus facilitating relief measures and rehabilitation of services.

Disaster preparedness is an ongoing, multisectoral activity. It forms an integral part of the national system responsible for developing plans and programs for disaster management (prevention, mitigation, preparedness, response, rehabilitation, or reconstruction). The system, known by a variety of names depending on the country, depends on the coordination of a variety of sectors to carry out the following tasks:

- Evaluate the risk of the country or particular region to disasters;
- Adopt standards and regulations;
- Organize communication, information, and warning systems;
- Ensure coordination and response mechanisms;
- Adopt measures to ensure that financial and other resources are available for increased readiness and can be mobilized in disaster situations;
- Develop public education programs;
- Coordinate information sessions with news media; and
- Organize disaster simulation exercises that test response mechanisms.

PREPAREDNESS IN THE HEALTH SECTOR

As outlined in Chapter 2, the health sector forms an essential part of the intersectoral (national, regional, or local) system for disaster preparedness and response. Its organization and response mechanisms need careful planning, and should take into account the vulnerability of the country or a specific region, health policies and legislation on disasters, and the administrative and technical organization of the health sector's institutions. This includes coordination of mechanisms, development of technical plans and programs, training and research, and logistical and financial support.

RISK ANALYSIS AND DEVELOPMENT OF REALISTIC SCENARIOS

The health sector must have a clear understanding of the risk of the country or a particular region to potential major hazards, whether the cause is natural (geologic or hydrometeorological events), technological (chemical or radiological accidents), social (violence, war, or subversion), or biological (large epidemics). Hazard analysis is carried out by governmental and/or private institutions and requires knowledge in such areas as seismology, volcanology, meteorology, structural engineering, and epidemiology.

Vulnerability to disease outbreaks should be evaluated, and data obtained on housing, living conditions, overcrowding, basic sanitation, and antecedents or history of endemic or natural foci of disease.

The health sector is responsible for using the data provided by specialized institutions to determine the vulnerability of its essential facilities (hospitals, health centers, and administrative buildings), and lifelines that guarantee the operation of these institutions, such as water service, power, communications, and transportation, and its own response capabilities and mechanisms. When analyzing vulnerability, organizational as well as physical weaknesses should be assessed in order to develop realistic plans for health scenarios following disasters.

The first step in evaluating risk is to estimate the probability of hazards occurring. It is important, when possible, to obtain multi-hazard maps (usually available from the scientific community, industry, the press, political authorities, and other sources) or to create them. The second step is estimate the vulnerability for each region or area. These data will be collected from the national disaster management agency and other entities and in consultation with engineers, architects, planners, civil defense staff, and others.

Some countries are developing geographical information systems (GIS) that can be of great assistance in estimating levels of risk. They are generally located in institutions outside of the health sector, but their synthesis of information is useful for all sectors and activities of the country. They are commonly used for development and planning processes, which includes disaster mitigation.

HEALTH POLICY AND LEGISLATION ON DISASTERS

While health institutions can develop individual disaster preparedness plans, it is desirable for countries to have a clear policy on disaster prevention and management. Legislation should require health institutions to develop preparedness and response plans, to institutionalize the plans as part of their normal activities, to use simulations to test the plans, and to assign financial resources for their development and maintenance. Hospital disaster plans should be required for a hospital's accreditation.

PREPARATION OF DISASTER PLANS

The following guidelines should be kept in mind when preparing health sector disaster plans:

1. Identify probable health scenarios based on the hazard and vulnerability analysis, and use this knowledge as a basis for creating a disaster plan. Decisions have to be made as to the resources that should be mobilized in planning for the most probable scenario as opposed to the "worst case scenario" (which is unlikely to occur in a lifetime).

2. List all probable events and likely health needs created by different scenarios. To be effective, planning must be directed toward specific and realistic objectives, such as how to cope with unsolicited assistance or how best to use available resources.

3. Plan for the main features of administrative response, such as the location and general responsibilities of key officials. Do not complicate plans with detail. Allow for ad hoc and improvised responses to fill in gaps.

4. Subdivide plans into self-sufficient units. Adequate response to a disaster does not usually require specialized staff (e.g., hospital administrators) to be familiar with all aspects of the plan.

5. Disseminate the plan widely. People with roles to play in the disaster plan must be very familiar with it, which demands considerable training. Many good plans have failed during emergencies because of inadequate dissemination and practice.

6. Include exercises to test the plan periodically. Plans are not realistic if they are not tested. The absence of actual testing will largely negate even the best of abstract plans.

7. Include systems for early warning and information so that the public can adopt self-protection measures or reach temporary shelters if evacuation is necessary. Public information should come from authoritative and competent sources and have well defined formats so that messages are clear and precise. Warning systems for different types of disasters should be standardized at the national level and tested during simulations. The public must be aware of how warning systems work prior to the onset of an event.

8. Compile an information package with basic demographic information, including epidemiological data. The package should include topographical maps showing roads, bridges, and rail lines; the location and basic layout of health facilities; and other information that would assist in response. The package should be stored so that it can be rapidly retrieved in case of disaster. Where Geographic Information Systems (GIS) exist, they can be very useful; when they are maintained by other ministries or sectors, they can be shared.

COORDINATION MECHANISMS

If the health sector's disaster preparedness plan is to be successful, clear mechanisms for coordinating activities with other sectors and internationally must be in place.

The Health Disaster Coordinator is in charge of preparedness activities and coordinating plans with government agencies, including civil protection, armed forces, and foreign relations divisions; United Nations and other international agencies; Red Cross and other NGOs; and entities responsible for housing, communication, power, and water services. It is particularly important for the Health Disaster Coordinator to maintain ongoing communication and coordination with civil protection agencies and the PAHO/WHO Emergency Preparedness Program in each country.[1]

---

[1] For more information visit the PAHO Web site (http://www.paho.org/english/ped).

RELATIONS WITH THE MEDIA

The media play an important role in providing critical information to the affected population and the national and international audience in the event of a disaster. It is essential that authorities and media practitioners share an understanding of the objectives of information dissemination, as well as their respective roles in the disaster. Ongoing meetings or seminars between members of the media and disaster managers to clarify these roles and responsibilities are strongly recommended as part of disaster planning.

The media also play an essential role in educating the community about simple but critical measures that can be adopted to lessen the effects of disaster. The health sector should use the media to convey such messages on disaster preparedness and mitigation.

TECHNICAL HEALTH PROGRAMS

In the event of disaster, the health sector is responsible for treatment of casualties, epidemiologic surveillance and disease control, basic sanitation and sanitary engineering, oversight of health care in camps or temporary settlements for displaced persons and refugees, training, and logistic resources and support.

The responsibilities of the health sector in the aftermath of a disaster cover practically every aspect of normal pre-disaster operations. No technical department or support service can remain uninvolved or immobilized in case of a major disaster. Preparedness should address all health activities and disciplines and cannot be limited to the most visible aspects of mass casualty management and emergency medical care. To reinforce these responsibilities, a standing advisory committee (see Chapter 2) comprising specialists from health disciplines should meet on a regular basis to review preparedness activities and disaster plans in their respective areas of operation.

Treatment of Casualties

Prehospital and hospital plans for treating casualties are essential in organizing health services for disaster situations. The prehospital disaster plan focuses on search and rescue of victims requiring either specialized medical personnel or equipment, as in the case of persons trapped in buildings collapsed by earthquakes. Reliance on external assistance for search and rescue (SAR) activities should be minimized; instead, the health sector should promote the development of a national search and rescue capacity familiar with modern techniques and equipment.

Other prehospital activities include: first aid administered at the disaster site and, depending on the severity of injury, providing immediate treatment. The injured are identified or tagged at the disaster site, and classified according to priority for treatment and/or transfer to hospital. This process, known as triage, uses an internationally accepted color coding system (see Chapter 6). Because many health workers are unfamiliar with mass casualty management, it should be included in the medical and paramedical curricula in health schools.

The hospital disaster plan refers to the organization within a hospital, and focuses on: development of emergency plans, training, information, safety of patients and hospital personnel, evacuation, and availability of medicines and medical supplies for emergency treatment. The plan also addresses backup systems for communication, power, water supply, and transportation. It should form part of the hospital disaster response network, with clear procedures for patient referral and transport.

Identification of Bodies

Identification of bodies requires careful coordination with forensic medicine departments. The health sector should develop protocols for the identification and conservation of cadavers, death certification, and local and international transport, as necessary. Not all countries will find it practical to maintain expertise in this field, but health authorities should be familiar with the approach and establish contacts with potential sources of technical cooperation.

Epidemiological Surveillance and Disease Control

As discussed in Chapter 7, the type of disaster determines the levels of morbidity and mortality in a population. However, as part of the epidemiological surveillance system, it is advisable to institute warning mechanisms with a list of potential illnesses related to the type of disaster, establish a simple data collection system, and set up special programs such as those for vector control or control of diarrheal diseases or nutritional problems. This is not to be improvised. It is the responsibility of the epidemiology department to prepare itself and the health services to face this challenge at the time of crisis.

Technological accidents require a specialized surveillance system. Disaster planning should include prior designation of information centers and treatment for chemical poisoning and for exposure to ionizing radiation. Although not addressed in this publication, health workers must acquire special skills to respond to technological accidents.

Many different resources are available to support this training.[2] In addition to printed and audiovisual materials, an increasing body of work is available for consultation via the Internet.

Basic Sanitation and Sanitary Engineering

Basic sanitation and sanitary engineering include water supply and wastewater disposal, solid waste disposal, food handling, vector control, and home sanitation (see Chapter 8). The environmental health department and the water authorities should collaborate in developing contingency plans to ensure that these vital services are uninterrupted regardless of the magnitude of the disaster.

---

[2] Among other documentation centers, the Regional Disaster Information Center (CRID), a multi-agency center based in San José, Costa Rica, collects and distributes documentation relating to various aspects of disasters and disaster management. The CRID database is accessible through the Internet (http://www.disaster.info.desastres/net/CRID).

Health Management in Shelters or Temporary Settlements

The health sector is responsible for establishing basic health programs for temporary shelters, including a surveillance and control system for infectious diseases and nutritional surveillance. Children should receive appropriate vaccinations, and opportunities should be taken to provide basic health education to residents of temporary settlements.

Training Health Personnel and the Public

Health ministries in countries vulnerable to disasters should institute comprehensive in-service training programs. Specific training in first aid, search and rescue (SAR) techniques, and public hygiene for the population at risk should be given, and health officials should receive ongoing instruction in disaster management issues in their respective areas of responsibility. Health institutions should recruit professional staff with qualifications in disaster management to be in charge of disaster programs.

It is even more important, perhaps, for professional training institutions (universities, schools, etc.) to include disaster preparedness and response in their regular curricula or as part of continuing education programs.[3] The health sector should also encourage the development of research protocols to be applied during the disaster phase to identify factors that would contribute to improving disaster management, or to characterize the effects of a disaster on the health of the population.

Logistical Resources and Support

The health sector must have a budget for preparedness as well as disaster response activities. Mechanisms should be in place to allow for the quick mobilization of resources after a disaster, rather than using normal administrative procedures that are generally bureaucratic and time-consuming.

It is usually uneconomical for individual health facilities, particularly hospitals, to stockpile disaster relief supplies. Medicines with expiration dates, for example, should not be kept in large quantities. As part of preparedness planning, hospitals should join a network of national or regional institutions that maintain stocks that can be quickly distributed. These might include stocks in government or military warehouses. Chapter 12 outlines factors to consider in managing the receipt, inventory, and distribution of humanitarian supplies.

Simulation Exercises

Simulations should take place with the participation of health authorities and operative personnel. They are the only way to keep plans up to date, especially during prolonged periods when emergencies do not occur. There are a variety of techniques for conducting simulation exercises:

---

[3] This is taking place in several Central American training institutions, where the modular approach has been quite successful. These training activities have received the technical support from the WHO/PAHO Collaborating Center at the Universidad de Antioquia in Medellín, Colombia, which has strengthened links worldwide.

- Desktop simulation exercises (sometimes called "war games" in military jargon) use paper or computer-based scenarios to improve coordination and information sharing and test the decision-making process.
- Field exercises are more costly, but are highly visible and are popular because they actually test the activation of a disaster plan in simulated field conditions. While these exercises cannot realistically reproduce the dynamic and chaos of real life disasters, they are very useful when intended to detect the inevitable errors, lack of coordination, or deficiencies of the simulated response. A critical evaluation is the essential conclusion of these exercises. A perfect field exercise is one that exposes many shortcomings in the disaster plan.
- Drills are designed to impart specific skills to technical personnel (e.g., search and rescue, ambulance, firefighting personnel). A perfect drill is one that leads to a flawless repetition of the intended task under any circumstance.

CHAPTER 4.
DISASTER MITIGATION IN THE HEALTH SECTOR

It is virtually impossible to prevent the occurrence of most natural hazards, but it *is* possible to minimize or mitigate their damaging effects. In most cases, mitigation measures aim to reduce the vulnerability of the system (for example, by improving and enforcing building codes). In some cases, however, mitigation measures attempt to reduce the magnitude of the hazard (e.g., by diverting the flow of a river). Disaster prevention implies that it is possible to completely eliminate the damage from a hazard, but that is still not realistic for most hazards. An example would be that of relocating a population from a floodplain to an area where flooding has not occurred or is unlikely to occur. In such a case, the vulnerability will be brought to zero, since from a public health or social point of view there is no vulnerability where there is no population.

Medical casualties could be drastically reduced by improving the structural quality of houses, schools, and other public or private buildings. Although mitigation in these sectors has clear health implications, the direct responsibility of the health sector is limited to ensuring the safety of health facilities and public health services, including water supply and sewerage systems.

In the last two decades in Latin America and the Caribbean, nearly 100 hospitals and more than 500 health centers have suffered damage as a result of hazards. In the worst cases, hospitals collapsed, killing patients and medical staff. More commonly, services to the community were interrupted, jeopardizing the health of the population. In many instances, even years after an event, repairs have not been completed. When water supplies are interrupted or contaminated, public health consequences can be severe. In addition to the social costs of such damage, the costs of rehabilitation and reconstruction severely strain economies.

HEALTH SECTOR DISASTER MITIGATION PROGRAMS

Because of the variety and cost of mitigation activities, priorities for implementing these measures must be established. In the health sector, this is the function of the national health disaster management program, working with experts in such areas as health and public policy, public health, hospital administration, water systems, engineering, architecture, planning, education, etc. A specialized unit within the national health disaster management program should coordinate the work of these professionals. Mitigation complements the disaster preparedness and disaster response activities of the program.

The mitigation program will direct the following activities:

1. Identify areas exposed to natural hazards with the support of specialized institutions (meteorology, seismology, etc.) and determine the vulnerability of key health facilities and water systems.
2. Coordinate the work of multidisciplinary teams in developing design and building codes that will protect the health infrastructure and water distribution from damage in the event of disaster. Hospital design and building standards are more stringent than those for other buildings, since hospitals not only protect the well-being of their occupants, but must remain operational to attend to disaster victims.
3. Include disaster mitigation measures in health sector policy and in the planning and development of new facilities. Disaster reduction measures should be included when choosing the site, construction materials, equipment, and type of administration and maintenance at the facility.
4. Identify the priority hospitals and critical health facilities that will undergo progressive surveys and retrofitting to bring them into compliance with current building standards and codes. The function of a facility is an important factor in establishing its priority. For example, in earthquake zones, a hospital with emergency medical capacity will have higher priority in the post-disaster phase than a facility that treats outpatients or those who could be quickly evacuated. Create mitigation committees at the local level to identify key facilities and ensure that mitigation measures are implemented in all projects.
5. Ensure that disaster mitigation measures are taken into account in a facility's maintenance plans, structural modifications, and functional aspects. In some cases, the facility may be well designed but successive adaptations and lack of maintenance increase its vulnerability.
6. Inform, sensitize, and train those personnel who are involved in planning, administration, operation, maintenance, and use of facilities about disaster mitigation, so that these practices can be integrated into their activities.
7. Promote the inclusion of disaster mitigation in the curricula of professional training institutions related to the construction, maintenance, administration, financing, and planning of health facilities and water distribution systems.

Annex I describes the steps involved in establishing a national disaster mitigation plan for hospitals in an earthquake-prone region.

VULNERABILITY ANALYSIS IN HEALTH FACILITIES

The first phase of the disaster mitigation program is to conduct a vulnerability analysis, i.e., to identify weaknesses in the system that may be exposed to hazards. Since the objective of this analysis is to establish priorities for either retrofitting or repair, there is no reason to perform the study if there is no intention of implementing the recommended mitigation measures.

A multidisciplinary team (composed of health administrators and specialists in natural hazard assessment, environmental health, engineering, architecture, planning, etc.) conducts the vulnerability analysis. The team will identify potential hazards, classify the location of the system (soil quality, access routes, etc.), determine the expected performance of the system, and analyze maintenance operations. The team will then be in a position to present the results of this initial, low-cost study to the "owner" or "client" and propose mitigation measures, taking into account political willingness and financial constraints. Based on the decision taken, a quantitative vulnerability analysis study is then performed.

Professionals with expertise in natural hazard evaluation, methods of risk analysis, and conducting retrofitting projects generally are hired from outside of the hospital or water system being targeted. Training should take place during the analysis, so that institutions gain the basic capacity to lessen their vulnerability.

Vulnerability analysis must take place regularly, as both hazards and vulnerability change over time.

DISASTER MITIGATION IN HEALTH FACILITIES

Building standards for health facilities are different than those for most buildings, particularly those health facilities that will be under increased pressure to attend to medical emergencies in a disaster's aftermath. Mitigation measures in hospitals have to be oriented, first, to avoiding loss of life of patients and staff, and second, to ensuring that the hospital will function properly after the hazard's impact. Each component of the hospital must undergo vulnerability analysis.

The following factors are considered when conducting vulnerability analysis and preparing mitigation plans for medical facilities:

1. Structural elements, which include a building's load-bearing components, such as beams, supporting columns, and walls;
2. Nonstructural elements, including architectural elements (exterior non–load-bearing walls, in-fill walls, partition systems, windows, lighting fixtures, and ceilings); lifeline systems (water, power, and communication systems); and the building's contents (medicines, supplies, equipment, and furnishings). Nonstructural damage can be severe, even if the building structure remains intact;
3. Functional elements, which include the physical design (site, external and internal distribution of space, access routes), maintenance, and administration. The administrative and operational aspects of the facility (including disaster plans and performance of simulation exercises) are addressed as part of preparedness activities.

The analysis of structural components should be carried out first, since these results are used to determine the vulnerability of nonstructural and functional elements.

Once a facility's weaknesses are identified, a mitigation plan can be developed. Considering the costs and technical complexity of different measures, it is quite legitimate to begin with the least expensive measures. If resources permit, the structural components, which generally are the most complex and require substantial investment, will be retrofitted. The cost of applying seismic-resistant measures to existing structures ranges between an estimated 4% to 8% of the total cost of the hospital. In the case of mitigation measures for structures exposed to hurricanes, the percentage is even less.

Functional elements, while requiring only modest capital investment, may be surprisingly complex and time consuming. In situations where there are severe political or financial obstacles to undertaking mitigation projects, the application of simple, low-cost measures, such as those applied to nonstructural elements, will reduce the probability of failure of systems in the event of small-scale hazards, which occur most often. The role of maintenance engineers is important in such cases.

All parties concerned (the clients or owner of the institution, financial officers, and technical personnel) should discuss the decision to undertake a mitigation program at the national or local level. Where there are limited economic and technical resources, the mitigation plan should be programmed for completion over a period of several years.

DISASTER MITIGATION IN DRINKING WATER SUPPLY AND SEWERAGE SYSTEMS

Drinking water supply and sewerage systems in urban and rural areas are particularly vulnerable to natural hazards. The systems are extensive and often in disrepair. When water supply is contaminated as a result of disasters, the population is at increased risk of contracting disease, and sanitation quickly deteriorates. Indirect health consequences are often difficult to evaluate and the costs to repair the system are generally very high. For example, as a result of the Mexico City earthquake in 1985, an estimated 37% of the city's population did not have access to water in the weeks following the disaster. As a result of the effects of the El Niño phenomenon in 1997–1998, the population of Manta, Ecuador, went without water for three months. Costs to repair the damaged infrastructure in this case exceeded US$ 600,000; losses to the water authority due to uncollected receipts exceeded

TABLE 4.1. Hospitals and health centers damaged or destroyed, by selected natural disasters, Latin America and the Caribbean.

Disaster	Hospitals and health centers damaged	Beds out of service
Earthquake, Chile, March 1985	79	3,271
Earthquake, Mexico, September 1985	13	4,387
Earthquake, El Salvador, October 1986	7	1,860
Hurricane Gilbert, Jamaica, September 1988	24	5,085
Hurricane Joan, Costa Rica and Nicaragua, October 1988	4	...
Hurricane Georges, Saint Kitts, September 1998[a]	1	170
Hurricane Georges, Dominican Republic, September 1998	87	...
El Niño, Peru, 1997–1998	437	...
Hurricane Mitch, Honduras, November 1998	78	...
Hurricane Mitch, Nicaragua, November 1998	108	...

[a]In the 35 years that the Joseph N. France Hospital in Saint Kitts has operated, it has been seriously damaged by hurricanes on 10 occasions.

— Not available.

US$ 700,000. Costs to repair damage to the aqueduct system resulting from the Limón, Costa Rica, earthquake in 1992 exceeded US$ 9,000,000.[1]

Authorities that operate and maintain water systems should have strategies directed at reducing these systems' vulnerability to natural hazards and procedures to quickly and effectively restore services in the event of a disaster. As with health facilities, vulnerability analysis is the first step in identifying and quantifying the effect of potential hazards on the performance and components of the system. This process is complicated by the fact that drinking water and sewerage systems are spread over large areas, composed of a variety of materials, and exposed to different types of hazards, including landslides, flooding, strong winds, volcanic eruptions, or earthquakes.

The analysis of the water and sewerage system is conducted by a team of professionals with expertise in natural hazard assessment, environmental health, and civil engineering, along with water service company personnel who are familiar with service operation and maintenance. Their focus is on operation and maintenance, administration, and potential impacts on service, as outlined below:

- **Operation and Maintenance.** The team analyzes how the overall system performs. Important factors for drinking water are the capacity of the system, the amount supplied, continuity of service, and quality of water. For the sewerage system, coverage, drainage capacity, and quality of effluents are evaluated. Information on the vulnerability of specific components (intakes, pipelines, treatment plants, storage tanks, drainage systems, etc.) indicates how the failure of one component will affect overall performance.
- **Administration.** The team ascertains the ability of the water service company to provide effective response by reviewing its disaster preparedness, response, and mitigation program. This includes mechanisms to disperse funds in emergency situations and necessary logistical support (personnel, transportation, and equipment) to restore water service. The analysis reveals whether disaster mitigation measures are included in routine maintenance, if necessary equipment and replacement parts are available for emergency repairs, and staff are trained in disaster response.
- **Impact on Service.** The team analyzes the potential impact of different hazards on specific components. Special attention is given to the location of a component and risks in the area, its condition (for instance, corrosion in pipes), and how critical it is to overall performance of the system. The team also estimates the time required to make repairs, the potential number of broken connections, and decreases in water quality or quantity that would result in rationing.

This information is used in the disaster preparedness plan to indicate the need to provide alternative water sources, the amount of time required to restore water service, and connections and installations that have priority for special monitoring, repair, or replacement.

---

[1] Pan American Health Organization. Centro Panamericano de Ingeniería Sanitaria y Ciencias del Ambiente (Publicación N°. 96.23). Estudio de caso: terremoto del 22 de abril de 1991, Limón, Costa Rica. Lima: OPS/CEPIS; 1996.

Mitigation measures for water systems include retrofitting, replacement, repair, placement of back-up equipment, and improved access. The mitigation plan may recommend such measures as relocation of components (as in pipelines or structures located in unstable terrain or close to waterways), construction of retaining walls around installations, replacement of rigid joints, and use of flexible piping.

Applying mitigation measures to existing systems is complex and costly. Water authorities, administrators, and operators must take responsibility for ensuring that disaster mitigation measures form part of the design and routine operation of these systems, and are included in the master plan and execution of any expansion to the system.

CHAPTER 5.
COORDINATION OF DISASTER RESPONSE ACTIVITIES AND ASSESSMENT OF HEALTH NEEDS

The response to disasters, both by nations affected and from the international community, has gradually improved in Latin America and the Caribbean in the last 30 years. With the evolution of national disaster management agencies, disaster response by governmental and nongovernmental institutions is better coordinated and based on pretested advance plans.

NATIONAL EMERGENCY COMMITTEE

After a disaster, all resources of the affected country are mobilized. Ideally, they are placed under the direction of a single national authority in the National Emergency Committee, in accordance with emergency legislation adopted beforehand. This Committee should be attached to the national disaster management agency and assumes overall disaster response coordination from a designated Emergency Operations Center. The National Emergency Committee is chaired by the President of the country or by his/her representative. Where there is an impact on the health situation of the population, the Health Disaster Coordinator will provide the link between the overall national disaster management authority and the health sector. The Minister of Health or his/her representative is the official health representative on the National Emergency Committee.

Membership of the National Emergency Committee will vary depending on the nature of the disaster. For example, its composition during a complex emergency would differ from that during a cholera epidemic. Figure 5.1 illustrates a proposed organization for the Committee. The organization will reflect each country's specific administrative, social, and political structure. Final responsibility for equipment such as heavy vehicles and telecommunications, the authority to request or accept external assistance, and clearance to issue news releases on health matters will probably lie outside the health sector.

In support of this national structure, the United Nations system in each country has established a Disaster Management Team. That team is chaired by the U.N. Resident Representative and is composed of the heads of U.N. agencies present in the country and in some cases of major bilateral agencies and NGOs. Chapter 13 discusses the role of international agencies in humanitarian assistance.

FIGURE 5.1. Members of the National Emergency Committee.

			National disaster coordinator			
Head of communications (Public relations officer, public information, radio communications)	Head of damage and needs assessment	Head of health and welfare	Head of operations (e.g., transport)	Head of security	Head of policy-making (representatives of executive or legislative branch)	Other

Note: These members are usually permanently stationed in the Emergency Operations Center during the disaster response phase. Membership is according to the hazard. Other officials should be invited for special briefing sessions.

HEALTH EMERGENCY COMMITTEE

In case of disaster, the major function of the Health Disaster Coordinator is to advise, or execute on behalf of the health sector authority (e.g., Minister of Health) operational coordination and to mobilize all possible health resources to save lives and limit material losses to the health sector.

In support of these activities, a Health Emergency Committee is convened. This Committee will include senior representatives of the health ministry, sanitation and water services, major accredited voluntary agencies, and other ministries involved in health relief programs. In contrast to the standing advisory committee for disaster preparedness mentioned, which has a large membership, the size of the Health Emergency Committee should be limited. Meetings that include too many staff members have impeded quick and efficient decision-making in several disaster situations.

A press or communications officer should be attached to the Health Emergency Committee to disseminate information and decisions (see Chapter 3).

Figure 5.2 illustrates the functional areas that the Health Disaster Coordinator and Health Emergency Committee should consider in organizing humanitarian operations. Several activities, such as transportation, supply management, and volunteer coordination, must be integrated with the corresponding areas of the National Emergency Committee (Figure 5.1). The health transportation unit, for example, will work closely with and under the direction of the National Emergency Committee's transport section.

Should the creation of the National Health Disaster Management Program have been overlooked, a senior health official must be appointed in the immediate post-disaster phase to represent the health sector on the National Emergency Committee. His or her tasks, with the support of the Health Emergency Committee, will be to direct the sector's relief activities and set its priorities, clear news releases, approve requests for external cooperation, and accept or reject offers of assistance on behalf of the Minister of Health.

ASSESSMENT OF NEEDS

The major administrative problem in many relief operations is the mass of conflicting and often exaggerated reports about the extent and effects of the disaster.

FIGURE 5.2. Coordination of Health Emergency Activities.

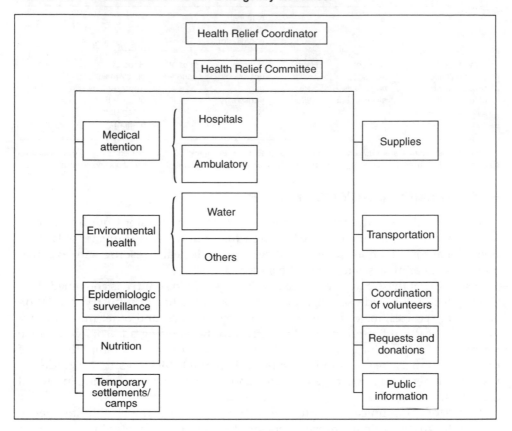

Factual information is necessary to meet three main objectives: to define the affected population; identify and anticipate its unmet needs by assessing the extent of damage and existing local human and material resources; and identify potential secondary risks to health. The Health Disaster Coordinator will also require information in order to keep the international assistance community abreast of the changing situation so that it can respond appropriately; provide verified facts to the national and international media in order to avoid unsubstantiated reports, such as of disease outbreaks, that may provoke inappropriate responses; and keep the local population accurately informed about available services and prevent or counteract rumors.

Timing of the information is usually more important that its completeness and accuracy, as decisions need to be made as soon as possible in the emergency phase, with the data that are available. Within the first few hours of the hazard impact, authorities must have an idea of the overall extent of the disaster, allowing them to take the first decisions for the general population affected. Subsequently, data will be progressively adapted to a smaller scale, culminating, if possible, in the satisfaction of individual needs.

Information Requirements

Figures 5.3 and 5.4 show probable changes in needs and relief priorities at different periods after earthquakes and floods, respectively. The major information requirements for emergency relief after different types of disaster are: (1) the geographic area affected, an estimate of the population size, and its location in the affected area; (2) the status of transport (rail, road, air) and communication systems; (3) the availability of potable water, food stocks, sanitary facilities, and shelter; (4) the number of casualties; (5) damage sustained by hospitals and other health facilities in the affected area, their capacity to provide services, and their specific drug, equipment, and personnel needs; (6) the location and numbers of people who have moved away from their homes (for example, into urban areas, roadsides, or high ground); and (7) estimates of the numbers of dead and missing. The last has low priority when the major concern is to provide essential services to survivors.

In the first few days, the provision of immediate humanitarian assistance and the collection of information will be conducted simultaneously. As urgent relief needs are met, information can be collected on specific topics to define further priorities.

Background Information

Collecting and interpreting information will be simplified if background information is maintained in a summarized and easily accessible form (displayed, when possible, on maps) as part of a pre-disaster plan. It should show the size and distribution of the population in the area; major communication lines and topography; distribution and services provided by health facilities, with notations of those that might be particularly vulnerable to natural disasters as determined by prior vulnerability analysis; location of large quantities of food, medicine, and health supplies in government stores, commercial warehouses, and those of major voluntary

FIGURE 5.3. Changing needs and priorities following earthquakes.

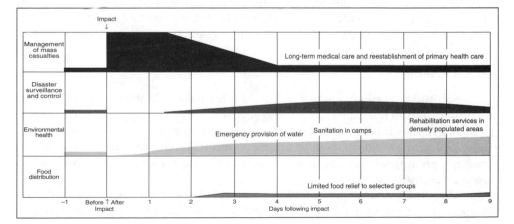

FIGURE 5.4. Changing needs and priorities following floods/sea surges.

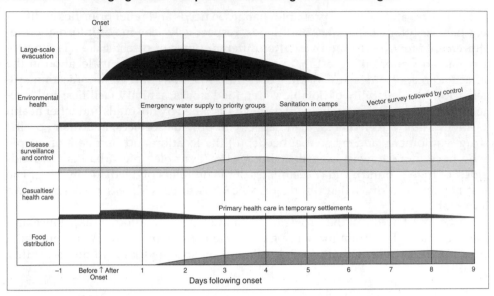

and international agencies; key people and organizations active in relief; and the location of potential evacuation areas.

Computerized geographical information systems offer a very promising tool for storing and displaying these data. They require a significant and sustained commitment of human resources to maintain, however, and it is difficult to justify their use exclusively for disaster response. As mentioned earlier, they are most commonly used for a country's development and planning processes. Where GIS exist, they should include the essential information to help authorities make decisions in a post-disaster situation.

METHODS OF GATHERING INFORMATION AFTER HAZARD IMPACT

Information can be obtained in five main ways: from aerial observation (light aircraft, helicopters, satellites); reports directly from the community and relief workers; reports from the media; regular reporting systems; and surveys.

Aerial Observation

Low-altitude overflights may yield rapid information on the geographic extent of damage and major damage to bridges, roads, and other specific lines of communication. This information is of limited use in determining the operational capacity of facilities and damage to underground installations, however. Helicopters have great flexibility and health workers should try to use them early on in needs assessment.

Satellite imagery is rapidly replacing aerial observation. Although satellite images are extremely valuable in determining the extent of physical damage, they are, at present, of little use in identifying the needs for urgent medical care.

Reports from the Community and Humanitarian Workers

Reports are received from community leaders, administrators, and local authorities, but they often have serious information gaps, since they lack information about isolated, severely affected communities. The respondent may have little accurate information to report, and may exaggerate the importance or urgency of some needs. Where reasonable doubt exists, the health assessment committee should not accept requests for large-scale relief at face value, but should try to discover why a particular need is said to exist. Humanitarian teams sent to affected communities should also be instructed to provide basic information on health needs and the community's ability to cope with them.

Reports from the Media

The international and national media provide rapid reports on damage and health needs. Their technical relevance, accuracy, and completeness usually do not match their speed and coverage. An increasing number of decisions are based on those reports, however, as they are a valuable source of information for health authorities when planning and orienting their professional assessment of the situation.

Regular Reporting from Existing Facilities

Where communications can be reestablished rapidly, information must be sought directly from administrative centers, public and private hospitals, and other technical agencies about immediate medical care, water, food, and sanitation needs. As noted in Chapter 7 on disease surveillance, epidemiological techniques are particularly useful in gathering and evaluating this information.

If large numbers of casualties are expected, for instance, daily reports should begin to be gathered from major health facilities as soon as possible after the impact to determine their ability to cope with the increased load and their need for support. A standard reporting format should be used by all components of each agency (health ministry, social security agency, armed forces, NGOs, and the private sector). The information collected should include the number of casualties appearing for treatment each day, other patients, admissions, vacant beds, and deaths. If possible, attendance and admissions should be reported by broad age and diagnostic categories.

Essential material in short supply such as casting plaster or x-ray film and specific food, water, and power problems also should be reported.

Surveys

Objective and quantified information on certain health needs can be obtained only through systematic surveying. If existing information sources are inadequate or inaccurate, suitable surveys should be conducted as soon as possible. After a major disaster, surveys may be organized in the following three stages:

Stage 1

Within the first 24 to 48 hours, an initial rapid assessment of damage—called "quick and dirty"—generally conducted by helicopter and sometimes based on satellite imagery, delineates the affected area by examining all potentially affected areas. The physical condition of health, transport, and communications facilities, as well as the status of relief activities should be quickly assessed by gathering information provided by one or several of the above mentioned methods. This will be sufficient to establish the types of problems that have arisen, to serve as a basis for mobilizing specific relief, and to design more formal surveys. The initial survey is generally carried out with the assistance of the armed forces and the participation of international experts (e.g., PAHO/WHO and U.N. Disaster Assessment and Coordination Teams).

Familiarity with the area to be surveyed is most important. Participation by health professionals in the survey will be an asset, but is not essential, as the data are not highly technical and can be gathered by others.

There is generally a conflict between the need for assessing the overall problem and the urge to provide immediate humanitarian assistance. To resolve this, surveillance personnel should refrain whenever possible from giving medical care, and backup medical assistance must be provided.

Stage 2

During the second phase of assessment, which may vary from a few hours up to several days after the impact, a detailed multidisciplinary health survey must attempt to include all affected areas.

During the first days, a survey in outlying areas should include an assessment of the numbers of casualties and dead. A survey of health needs must be a part of emergency care so that the survey team can call in immediate medical backup. Information should be collected on: (1) the total number of casualties; (2) number requiring evacuation and their major diagnostic categories; (3) number requiring local treatment; (4) availability of essential health supplies and personnel; (5) continued aftercare likely to be needed for those receiving emergency treatment; and (6) need to supply or make temporary repairs to local medical facilities.

The detailed survey will also try to assess the immediate impact of the disaster on water quality and availability. The aim is to estimate the extent to which damage to water supply systems and other sanitation services immediately increases health hazards (e.g., transmission of diseases) when compared to pre-disaster conditions, not to assess their absolute quality.

The need for food, shelter, and protective clothing must also be assessed.

In contrast to the initial rapid survey, it is essential to have the most qualified available health professionals take part in this survey, since major humanitarian operations will be based on the findings. At least one survey team member should be chosen for his or her familiarity with local conditions. Since technical competence and prior experience in disaster assessment are major assets, regional or international personnel may have to be called on to provide expertise unavailable locally. Neighboring countries should consider pooling such resources before disasters occur, based on principles of technical cooperation among developing countries.

Transporting survey teams must be given highest priority, since other relief activities will be competing for available transportation. Specifically, survey team space should be sought on all relief transport if the teams do not have their own transportation. Helicopters are the most flexible and useful transport for such surveys.

Stage 3

In the third assessment stage, surveys of specific problems must be made. Damage to health facilities and related utilities should be surveyed throughout the affected area by competent technicians and engineers. These surveys will provide a basis for estimating reconstruction costs. If such cost estimates are not quickly available, scarce international relief funds cannot be suitably channeled to priority areas in the health sector. Finally, these surveys will start the continuing surveillance needed to direct health sector assistance activities rationally.

Too often, disaster managers have confused the assessment of emergency humanitarian needs with the evaluation of rehabilitation and reconstruction requirements. Humanitarian agencies and donors expect immediate data on the emergency needs and not on estimates of the long-term economic impact of damages and the cost of subsequent redevelopment. These data should be collected, but at a later stage.

CHAPTER 6.
MASS CASUALTY MANAGEMENT

Medical treatment for large numbers of casualties is likely to be needed only after certain types of disasters. Most injuries are sustained during impact, and, thus, the greatest need for emergency care occurs in the first few hours. Many lives have been lost because local resources have not been mobilized quickly.

The burden of organizing and delivering transport, first aid, medical care, and supplies falls on the affected country. Help from the international community is unlikely to make a difference in saving lives during the period of greatest need, because of the response time required.

In the classic care approach used most commonly to deal with a huge number of victims after a disaster, first responders are trained to provide victims with basic triage and field care before evacuating them to the nearest available receiving health care facility.

The management of mass casualties is divided into three main areas: prehospital emergency care (search and rescue, first aid, triage, and stabilization of victims); hospital reception and treatment; and redistribution of patients to other hospitals when necessary.

PREHOSPITAL EMERGENCY CARE

Search, Rescue, and First Aid

After a major disaster, the need for search, rescue, and first aid is likely to be so great that organized relief services will be unable to meet more than a small fraction of the demand. Most immediate help will come from uninjured survivors, and they will have to provide whatever assistance possible. Improvement in the quality and availability of immediate first aid services depends on increased training and preparation obtained through specialized agencies, for example, through courses taught to volunteers by fire brigades.

Field Care

Ideally, the transport of victims to the hospital should be staggered, and patients should receive adequate field treatment, allowing them to tolerate delays. However in reality, most injured persons will converge spontaneously on health facilities if they are at a reasonable distance, using whatever transport is available, regardless of the facility's operating status. Some victims may not request or be able to seek medical care, which makes active case finding an important part of any casualty relief effort. This is sufficient reason for creating mobile health-care teams to be deployed to the disaster site in addition to fixed first aid stations located near existing health facilities.

Providing proper treatment to casualties requires that health service resources be redirected to this new priority. Bed availability and surgical services must be maximized by selectively discharging routine inpatients, rescheduling non-priority admissions and surgery, and fully using available space and personnel. Certain physician responsibilities can be postponed and others can be delegated to health technicians, for example, treating minor wounds.

Provisions should be made for food and quarters for health personnel.

A center should be established to respond to inquiries from patients' relatives and friends; it should be staffed round-the-clock, using non-health personnel as necessary. The Red Cross may be well-equipped to direct this function.

Priority must be given to victim identification, which is becoming an increasingly specialized issue. Adequate mortuary space and services also must be provided.

Triage

When the quantity and severity of injuries overwhelms the operative capacity of health facilities, a different approach to medical treatment must be adopted. The principle of "first come, first treated," which is applied in routine medical care, is inadequate in mass emergencies. Triage consists of rapidly classifying the injured on the basis of the severity of their injuries and the likelihood of their survival with prompt medical intervention. It must be adapted to locally available skills. Higher priority is granted to victims whose immediate or long-term prognosis can be dramatically affected by simple intensive care. Moribund patients who require a great deal of attention (with questionable benefit) have the lowest priority. Triage is the only approach that can provide maximum benefit to the greatest number of injured in a disaster situation.

Although different triage systems have been adopted and are still in use in some countries, the most common classification uses the internationally accepted four-color code system. Red indicates high priority treatment or transfer, yellow signals medium priority, green is used for ambulatory patients, and black for dead or moribund patients.

Triage should be carried out at the disaster site in order to determine transportation priority and admission to the hospital or treatment center where the patient's needs and priority for medical care will be reassessed. Ideally, local health workers should be taught the principles of triage as part of disaster training to expedite the process when a disaster occurs. In the absence of adequately trained field health personnel, a triage officer and first aid workers must accompany all relief teams to the disaster site to make these assessments. Where an advanced medical post is established, medical triage will be conducted at the entrance to the post to determine the necessary level of care.

Persons with minor or moderate injuries should be treated near their own homes whenever possible to avoid social dislocation and the added drain on resources of transporting them to central facilities. The seriously injured should be transported to hospitals with specialized treatment facilities.

Tagging

All patients must be identified with tags stating their name, age, sex, place of origin, triage category, diagnosis, and initial treatment. Standardized tags must be chosen or designed in advance as part of the national disaster plan. Health personnel should be thoroughly familiar with their proper use.

HOSPITAL RECEPTION AND TREATMENT

At the hospital, triage should be the responsibility of a highly experienced clinician, as it may mean life or death for the patient, and will determine the priorities and activities of the entire staff.

Organizational Structure

Spirgi notes that effective management of mass casualties demands an organization of services that is quite different from that found in ordinary times. He states that a "hospital disaster plan designates the command structure to be adopted in case of disaster . . . [A] command team (consisting of senior officers in the medical, nursing, and administrative fields) . . . will direct people where to work according to the plan and mobilize additional staff and additional resources as required."[1]

Standardized Simple Therapeutic Procedures

Therapeutic procedures should be economical in terms of both human and material resources, and should be chosen accordingly. Health personnel and supplies should support these procedures. First line medical treatment should be simplified and aim to save lives and prevent major secondary complications or problems. Preparation and dissemination of standardized procedures, such as extensive debridement, delayed primary wound closure, or the use of splints instead of circular casts, can produce a marked decrease in mortality and long-term impairment.

Individuals with limited training can, in many instances, carry out simple procedures quickly and effectively. Certain more sophisticated techniques requiring highly trained individuals and complex equipment and many supplies (e.g., treatment of severe burns) are not a wise investment of resources in mass casualty management. This shift in thinking and action from ordinary practice to mass medical care is not easy to achieve for many physicians.

REDISTRIBUTION OF PATIENTS BETWEEN HOSPITALS

While health care facilities within a disaster area may be damaged and under pressure from mass casualties, those outside the area may be able to cope with a much larger workload or provide specialized medical services such as neurosurgery. Ideally, there will be a metropolitan system of emergency medical treatment

---

[1] Edwin H. Spirgi, *Disaster Management: Comprehensive Guidelines for Disaster Relief* (Bern: Hans Huber, 1979).

that allows hospitals to function as part of a referral network. At different levels of complexity, a network of prehospital relief teams can coordinate referrals from the disaster area. The decision to redistribute patients outside the disaster area should be carefully considered, since unplanned and possibly unnecessary evacuation may create more problems than it solves. Good administrative control must be maintained over any redistribution in order to restrict it to a limited number of patients in need of specialized care not available in the disaster area. Policies regarding evacuation should be standardized among all agencies providing relief in the disaster area, and hospitals that will receive patients.

The task of matching resources to needs is best accomplished by using a chart similar to that shown in Figure 6.1, which also can be enlarged and displayed on a wall. Hospitals are listed according to their geographic location, starting with those closest to the impact area. A visual display of the number of beds available, medical or nursing personnel required for round-the-clock services, shortages of essential medical items, and other needs will permit the Health Disaster Coordinator to direct external assistance to areas where needs and expected benefits are greatest. Patterns for redistributing resources or patients will emerge from analysis of the data. Such monitoring of hospital resources will be most useful when medical care is likely to be needed for an extended period.

If the Health Disaster Coordinator finds that the country's total health care capacity is insufficient to meet disaster-related needs, several alternatives must be considered. The best is rapid expansion of the country's own permanent facilities and staff, which has the advantage of fulfilling immediate needs and leaving behind permanent benefits. Another alternative, which has proved to be less desirable, may be staffed, self-sufficient, mobile emergency hospitals available from government, military, Red Cross, or private sources. If such a hospital is necessary, one from the disaster-affected country, or a neighboring country with the same language and culture should be considered first, and those from more distant countries considered second.

Foreign mobile hospitals may have several limitations. First, the time needed to establish a fully operational mobile hospital may be several days, while most casualties resulting from the immediate impact require treatment in the first 24 hours. Second, the cost of such a hospital, especially when airlifted, can be prohibitive and is often deducted from the total aid package given by the governmental or private relief source providing it. Third, such hospitals are often quite advanced technologically, which raises the expectations of the people they serve in a way that will be difficult if not impossible for local authorities to meet during the recovery period. Finally, it must be recognized that such hospitals are of great public relations value to the donor agency, which may inappropriately urge their use.

FIGURE 6.1. Monitoring of hospital resources.

1 NAME-PLACE	2 Specialty	3 BEDS		4 SURGEONS		5 ANESTHESIOLOGY		6 Other medical personnel required	7 Nurses required	8 Essential items in short supply	9 Other requirements or contracts
		a Total	b Available	a Present	b Required	a Present	b Required				
Hospital "A," Disaster City	General	850	8	5	4	5	4	2 pediatric 1 gynecol.	5	Suturing material, X ray film	Generator
Hospital "X," Normalville	Trauma-tology	450	145	5 (traumatol.)	—	3	—	1 gynecol.	1	Linen	Limited kitchen facilities

CHAPTER 7.
EPIDEMIOLOGIC SURVEILLANCE AND DISEASE CONTROL

RISK OF OUTBREAKS FOLLOWING DISASTERS

Natural disasters may increase the risk of preventable diseases due to adverse changes in the following areas:

1. **Population density.** Closer human contact in itself increases the potential spread of airborne diseases. This accounts in part for the reported increases in acute respiratory infections following disasters. In addition, available sanitation services are often inadequate to cope with sudden increases in populations.
2. **Population displacement.** The movement of disaster victims may lead to the introduction of communicable diseases to which either the migrant or indigenous populations are susceptible.
3. **Disruption and contamination of water supply and sanitation services.** Existing water supply and sewerage systems and power systems are particularly vulnerable, and may be damaged by natural disasters. In the aftermath of the 1985 earthquake in Mexico City, for example, millions of inhabitants remained without a piped water supply for as long as several weeks. Drinking water is prone to contamination caused by breaks in sewage lines and the presence of animal cadavers in water sources.
4. **Disruption of public health programs.** After a disaster, personnel and funds are usually diverted to relief. If public health programs (e.g., vector control programs or vaccination programs) are not maintained or restored as soon as possible, communicable disease transmission may increase in the unprotected population.
5. **Ecological changes that favor breeding of vectors.** Unusual periods of rain, with or without flooding, are likely to affect the vector population density. This may involve an increase in mosquito breeding sites or the introduction of rodents to flooded areas. This will be discussed in Chapter 8.
6. **Displacement of domestic and wild animals.** As with human populations, animal populations are often displaced as a result of natural disasters, carrying with them zoonoses that can be transmitted to humans as well as to other animals.
7. **Provision of emergency food, water, and shelter in disaster situations.** The basic needs of the population are often provided from new or different sources.

It is important to ensure that these new methods are safe and that they themselves are not the source of infectious disease.

Outbreaks of gastroenteritis, which are the most frequently reported diseases in the post-disaster period, are closely related to the first three factors mentioned above. Increased incidence (or at least increased reporting) of acute respiratory infections is also common in displaced populations. Vector-borne diseases will not appear immediately but may take several months to reach epidemic levels. It should be noted that humanitarian workers are at risk following sudden-impact natural disasters as well as the disaster victims.

The principles of preventing and controlling communicable diseases after a disaster are to:

- Implement as soon as possible all public health measures to reduce the risk of disease transmission;
- Organize a reliable disease reporting system to identify outbreaks and to promptly initiate control measures;
- Investigate all reports of disease outbreaks rapidly. Early clarification of the situation may prevent unnecessary dispersion of scarce resources and disruption of normal programs.

SETTING UP A DISEASE SURVEILLANCE SYSTEM

In emergency conditions, the routine disease surveillance system is either not up to the task, is disrupted as a direct consequence of the disaster, or cannot provide data quickly enough for timely decisions to be taken. It is recommended, therefore, that a local, syndrome-based surveillance system be prepared at the national level and temporarily instituted in the disaster aftermath. It should be a more flexible and faster reporting system than used in normal conditions. The routine surveillance system must be reestablished as soon as possible.

In order to collect and interpret data, it is essential that a national epidemiologist be assigned adequate epidemiologic and clerical staff who have transportation to the field and priority access to public or private laboratory facilities. In addition to the national epidemiologic staff, university departments, research centers, and bilateral or international agencies may provide trained epidemiologists and laboratory support nationally or regionally. The national epidemiologist should be the secretary of a disease surveillance and control subcommittee of the Health Emergency Committee (see Chapter 5). The subcommittee should provide direct feedback to hospitals and other health facilities where surveillance data are being collected.

The epidemiologist closest to the local reporting unit should investigate suspected disease outbreaks detected by the surveillance system as soon as possible. Until epidemiological assistance arrives, initial investigation and control measures are the responsibility of the local health unit.

Background data should be collected on the geographical areas affected, the major disease risks in the affected area (e.g., whether cholera or malaria are endemic), available resources, and the at-risk and affected populations. The national epidemiologist and Health Disaster Coordinator should designate syndromes or diseases to be included in the surveillance system (for example, fever, fever and diarrhea, fever and cough, trauma, burns, and measles). All health facilities and temporary shelters should institute the system, using a standardized form as shown in Figure 7.1.

In addition to information provided by the health system, information from humanitarian workers, NGOs, community groups and from unconventional sources such as newspaper accounts, including unconfirmed public rumors, are important as early warnings.

FIGURE 7.1. Post-disaster disease surveillance daily report.

Date .... Name of Reporter ....

From: ( ) Hospital ....
      ( ) Outpatient department ....
      ( ) Health center ....
      ( ) Clinic ....
      ( ) Others (Specify ....)

Locating Address	Telephone No.

Number of new cases with	Under 5 yrs.	Over 5 yrs.	Total
1. Fever (100°F or 38°C)			
2. Fever and cough			
3. Diarrhea with blood			
4. Fever and diarrhea			
5. Vomiting and/or diarrhea			
6. Fever and rash			
7. Dog bite			
8. Snake bite			
9. Burns			
10. Trauma			
11. Jaundice and diarrhea			
12. Deaths			
13. Other Specify:			

Comments: ....
....

Complete for evacuation centers only

No. of persons accommodated today ....
....

Report significant changes in water/sanitation/food supply ....
....

PRESENTATION AND INTERPRETATION OF COLLECTED DATA

Post-disaster surveillance is not designed to provide precise information on the incidence of a disease. However, it is important to have an *early warning* system that identifies when a given symptom complex or disease may be occurring in an affected area. This indication will provide the basis for more intensive investigation and, if necessary, lead to specific control measures. Where the affected population is well defined, as in camps for refugees or displaced persons, it will be both feasible and important for the national epidemiologist to determine rates and their change over time.

If the above-mentioned post-disaster surveillance system is effective, it will invariably result in an increase in the number of reported common and uncommon diseases and syndromes. This results from an increase in the number of reporting units, improved public awareness, and the greater concern and coverage by the mass media. This is not necessarily a reflection of increased disease, but rather the result of enhanced disease reporting compared with the pre-disaster pattern.

Negative reports are as important as positive ones, and each reporting unit should submit reports whether or not it has seen any disease ("zero reporting"). Negative reports will show that the unit is functioning and that health resources can be channeled elsewhere.

The epidemiologist closest to the local reporting unit should investigate suspected disease outbreaks detected by the surveillance system as soon as possible. Until epidemiological assistance arrives, initial investigation and control measures are the responsibility of the local health unit.

Summary reports of the surveillance system's technical findings should be fed back to the National Emergency Committee, hospitals, and health facilities and appropriate action taken to introduce necessary control measures if beyond the immediate competence of the epidemiologists (large sanitation programs, for example). The general public also should be informed of the risk of disease occurrence. The dilemma in some countries is whether an open policy of posting the available information on the Internet or elsewhere is in the best interest of public health. One school of thought is to limit dissemination to "validated data" approved by the health authorities. This approach does not take into consideration the need for rapid access to information and the fact that "invalidated" information will become public knowledge. A liberal policy encouraging NGOs and local authorities to exchange their observations and findings, electronically and otherwise, is to be encouraged. In all instances those reporting the data should state the source of their information.

The Health Disaster Coordinator should advise the national emergency committee on control measures to be taken to prevent the spread of disease.

LABORATORY SERVICES

Access to accurate and discrete rapid laboratory services is essential for public health management. It is important to establish the cause of any disease manifestation so that correct control measures can be taken. However, laboratory diagnosis is not required for subsequent patients who present the same symptoms. Laboratories must be able to diagnose diseases occurring locally, and be able to absorb an increase in samples when necessary. If access to a local laboratory cannot be guaranteed, reference laboratory assistance may be required.

Some diagnostic tests (ova and parasites in the stool, blood smear) can be made with a minimum of technology by field reporting units, but certain bacteriologic and virologic tests necessary for surveillance must be performed by referral laboratories. It is important to establish coordination with local, regional, national, or international laboratories to provide necessary diagnostic tests for disease surveillance and control. Because of difficulties in access to certain areas, it may be necessary to make special arrangements to transport specimens.

VACCINATION AND VACCINATION PROGRAMS

Special Programs

Health authorities are often under considerable public and political pressure to begin mass vaccination programs, usually against typhoid, cholera, and tetanus. This pressure may be increased by exaggerated reports of the risk of such diseases in the local or international press, and by the "offer" of vaccines from abroad.

Typhoid and Cholera

Rapidly improvised mass vaccination campaigns against typhoid and cholera should be avoided in Latin America and the Caribbean for several reasons:

1. The World Health Organization does not recommend typhoid and cholera vaccines for routine use in endemic areas. The newer typhoid and cholera vaccines have increased efficacy, but because they are multi-dose vaccines, compliance is likely to be poor. They have not yet been proven effective as a large-scale public health measure. In a disaster situation, vaccination might, however, be recommended for health workers. Good medical control must rely on effective case identification and treatment and effective environmental sanitation measures.
2. Vaccination programs require large numbers of workers who could be better employed elsewhere.
3. Supervision of sterilization and injection techniques may be impossible, resulting in more harm than good being done.
4. Mass vaccination programs may lead to a false sense of security about the risk of diseases and to the neglect of effective control measures.

Supplying safe drinking water and the proper disposal of excreta continue to be the most practical and effective strategy to prevent cholera and typhoid fever and should be given the highest priority after a disaster.

Tetanus

Significant increases in tetanus have not occurred after natural disasters. The mass vaccination of populations against tetanus is usually unnecessary. The best protection against tetanus is maintenance of a high level of immunity in the general population by routine vaccination before a disaster occurs, and adequate and early wound cleansing and treatment.

If tetanus immunization was received more than 5 years ago in a patient who has sustained an open wound, a tetanus toxoid booster is an effective preventive measure. In previously unimmunized injured patients, tetanus antitoxin should be administered only at the discretion of a physician.

Regular Programs

If routine vaccination programs are being conducted in camps or other densely populated areas with large numbers of children, it is prudent to include vaccination against tetanus, as indicated by public health guidelines, along with the other components of the vaccination program.

Measles, Polio, and Other Diseases Targeted for Eradication

Natural disasters may negatively affect the maintenance of ongoing national or regional eradication programs against measles and polio. Disruption of those programs should be closely monitored and prevented. Prevention and control programs for urban yellow fever, bubonic plague, or other vector-borne diseases should also be maintained to prevent the possible emergence or reemergence of diseases.

Vaccine Importation and Storage

Most vaccines—particularly measles vaccine—require refrigeration and careful handling if they are to remain effective. If cold-chain facilities are inadequate, they should be requested at the same time as the vaccines. Vaccine donors should ensure that adequate refrigeration facilities exist in the country before dispatching vaccines. During the emergency period it may be advisable for all imported vaccines, including those going to voluntary agencies, to be consigned to government stocks if cold-chain facilities are adequate.

The vaccination policy to be adopted should be decided at the national level only. Individual voluntary agencies should not decide to vaccinate on their own. Ideally, a national policy should be included in the disaster plan.

TRANSMISSION OF ZOONOSES

Displacement of domesticated and wild animals increases the risk of transmission of zoonoses, and veterinary and animal health services may be needed to evaluate such health risks. Epidemiologic identification/characterization of zoonoses is critical in evaluating the risks of occurrence of these diseases in areas affected by natural disasters. It is also essential to establish surveillance mechanisms to prevent human cases or outbreaks.

Dogs, cats, and other domestic animals frequently are taken by their owners to or near temporary shelters. Some of these animals are reservoirs of infections such as leptospirosis, rickettsioses, and bubonic plague, which can be transmitted through their excrement and urine or through ectoparasites, contaminating water and food.

Wild animals are reservoirs of infections that can be fatal to man. In searching for food and safety in the aftermath of a natural disaster, wild animals will come closer to affected communities, increasing the chance of transmission of illnesses such as hemorrhagic fever syndrome from the Hantavirus, hemmorhagic arboviruses, equine encephalitis, rabies, and infections still unknown in humans.

CHAPTER 8.
ENVIRONMENTAL HEALTH MANAGEMENT

PRIORITY AREAS FOR INTERVENTION

The continuation or quick rehabilitation of effective environmental health services is of primary importance in emergency health management after the onset of a natural disaster. First consideration should be given to areas where health risks have increased. These are areas with high population densities and severe disruption of services. Secondary priority areas are those with high population densities and moderate disruption, or those with moderate densities with severe disruption. Third priority should be given to areas with low population density and minor disruption of services.

Areas with high population densities are urban areas and their peripheries, camps for refugees and displaced persons, and temporary settlements. Hospitals and health clinics are among the facilities needing priority environmental health services.

Shelters are, by definition, short-term accommodations where the affected population can sit out the event—a hurricane, for example—and return to their homes as soon as possible. Such facilities are not designed to provide the required basic services for hundreds of people for prolonged periods. However, experience has shown that shelters remain occupied long after the event, preventing the resumption of the facility's normal operation.

Temporary camp settlements often create areas of extremely high population density where suitable services may be absent. Lack of water and basic sanitation facilities lowers the existing level of hygiene and increases the risk of communicable diseases. Diseases that are endemic in the areas of origin, transit, and settlement of displaced populations are of special concern. The International Federation of Red Cross and Red Crescent Societies (IFRC) reports that up to 50% of deaths among displaced people are caused by water-borne diseases. In selecting sites for temporary settlements, it is critical to ensure that the camp has access to a dependable water supply and other environmental health services.

Priority Environmental Health Services

Primary consideration should be given to services essential for protecting and ensuring the well-being of the people in high risk areas, with emphasis on prevention and control of communicable diseases. Post-disaster environmental health measures can be divided into two priorities:

1. Ensuring that there are adequate amounts of safe drinking water; basic sanitation facilities; disposal of excreta, wastewater, and solid wastes; and adequate shelter.
2. Providing food protection measures, establishing or continuing vector control measures, and promoting personal hygiene.

A checklist of possible disruptions in environmental health services is presented in Table 8.1.

The following actions are recommended to quickly re-establish adequate environmental health services and conditions:

1. Obtain information on population movements in or near stricken areas and map the location of camps for refugees and displaced persons, partially and/ or totally evacuated areas, relief worker settlements, and hospitals and other

TABLE 8.1. Natural disaster effects matrix.

Most common effects of specific events on environmental health		Earthquake	Hurricane	Flood	Tsunami	Volcanic eruption
Water supply and wastewater disposal	Damage to civil engineering structures	1	1	1	3	1
	Broken mains	1	2	2	1	1
	Damage to water sources	1	2	2	3	1
	Power outages	1	1	2	2	1
	Contamination (biological or chemical)	2	1	1	1	1
	Transportation failures	1	1	1	2	1
	Personnel shortages	1	2	2	3	1
	System overload (due to population shifts)	3	1	1	3	1
	Equipment, parts, and supply shortages	1	1	1	2	1
Solid waste handling	Damage to civil engineering structures	1	2	2	3	1
	Transportation failures	1	1	1	2	1
	Equipment shortages	1	1	1	2	1
	Personnel shortages	1	1	1	3	1
	Water, soil, and air pollution	1	1	1	2	1
Food handling	Spoilage of refrigerated foods	1	1	2	2	1
	Damage to food preparation facilities	1	1	2	3	1
	Transportation failures	1	1	1	2	1
	Power outages	1	1	1	3	1
	Flooding of facilities	3	1	1	1	3
	Contamination/degradation of relief supplies	2	1	1	2	1
Vector control	Proliferation of vector breeding sites	1	1	1	1	3
	Increase in human/vector contacts	1	1	1	2	1
	Disruption of vector-borne disease control programs	1	1	1	1	1
Home sanitation	Destruction or damage to structures	1	1	1	1	1
	Contamination of water and food	2	2	1	2	1
	Disruption of power, heating, fuel, water, or supply waste disposal services	1	1	1	2	1
	Overcrowding	3	3	3	3	2

1— Severe possible effect
2— Less severe possible effect
3— Least or no possible effect

medical facilities. This information will assist in determining which localities need priority attention.

2. Carry out rapid assessments to determine the extent of damage to the public water supply and waste disposal systems and the food production, storage, and distribution networks.
3. Determine the remaining operational capacity for delivering these basic environmental health services.
4. Make an inventory of available resources, including undamaged food stocks, human resources, and readily available equipment, materials, and supplies.
5. Determine the stricken population's immediate needs for water, basic sanitation, housing, and food.
6. Meet the needs of essential facilities as quickly as possible after basic human consumption needs are satisfied. Hospitals and other medical facilities may need increased water supplies if there are numerous casualties.
7. Ensure that refugees and displaced persons are properly housed and that the temporary settlements and other identified high risk areas have basic environmental health services.

For the efficient use of overburdened resources, it is important to immediately and accurately assess damages and identify needs for repair. Reports of damage and needs should include the following information:

1. Type, location, and extent of damage;
2. Accessibility and required means of transport to site of damage;
3. Remaining operational capacity;
4. Estimate of resources needed for repairs (personnel, equipment, and materials);
5. Estimated repair time.

Rapid assessment will assist in identifying resources required to restore the system immediately. If a list of needs is to be submitted to the donor community, it should be compiled quickly. Donor response is generally high in the days following a disaster, but soon subsides.

Human Resources

The unavailability of environmental health specialists will be a limiting factor when managing an emergency situation. Experts unfamiliar with local conditions and local environmental health services might misjudge priorities. First consideration should therefore be given to using locally available manpower. The local population should be actively encouraged to assist in providing needed resources and services. It should be clear that all immediate or short-term activities are directed to restoring pre-disaster services and not to making improvements beyond the pre-existing level. Nevertheless, the rehabilitation phase of the emergency provides an excellent opportunity to assess vulnerability of the water supply and sanitation system, and to carry out measures that will mitigate the effects of future events on water supply.

WATER SUPPLY

A survey of all public water supplies will have to be made, beginning with the distribution system and advancing to the water source. It is essential to determine physical integrity of system components, the remaining capacities, and bacteriological and chemical quality of the water supplied.

The main public safety aspect of water quality is microbial contamination. The first priority for ensuring water quality in emergency situations is chlorination; it is the best means for disinfection and emergency treatment of water because of its effectiveness, cost, and availability.

It is advisable to increase residual chlorine levels and raise water pressure as part of the relief operations. Low water pressure will increase the likelihood of infiltration of pollutants into water mains. Repaired mains, reservoirs, and other units require cleaning and disinfection.

A minimum free residual chlorine level of 0.7 mg/l is recommended in emergency situations. Routine testing of residual chlorine should start immediately with simple residual chlorine test kits and should continue well into the rehabilitation phase. In the absence of test kits, check if water has a distinct chlorine smell. Microbial contamination is likely if tests indicate the absence of residual chlorine in drinking water, unless bacteriological analyses prove otherwise. However, such analysis requires long periods of incubation (at least 8–24 hours), while residual chlorine levels can be measured in the field in a few minutes.

Chemical contamination and toxicity are a second concern in water quality and potential chemical contaminants have to be identified and analyzed. If there is justified concern that the water source is contaminated with toxic substances from a spill or heavy metals from volcanic activity, alternative water sources should be sought.

Alternative Water Sources

In general order of preference, consideration should be given to the following alternative water sources:

1. Deep groundwater;
2. Shallow groundwater and spring water;
3. Rain water;
4. Surface water.

Private water supply sources belonging to dairies, breweries, food and beverage plants, tourist resorts, and other industrial and agricultural developments often exist in the vicinity of a disaster stricken community. Pre-emergency arrangements with the owners of these systems will facilitate the use of the source in case of emergency.

Sources located near and/or downstream from sewage outfalls, chemical plants, abandoned or operational solid waste disposal sites, abandoned or operating mines, and any other hazardous sites should be considered suspect until an environmental health specialist familiar with the local conditions recommends otherwise.

Existing and new water sources require the following protection measures:

1. Restrict access by people and animals. If necessary, erect a fence and appoint a guard;
2. Ensure adequate excreta disposal at a safe distance from the water source;
3. Prohibit bathing, washing, and animal husbandry upstream of intake points in rivers and streams;
4. Upgrade wells to ensure they are protected from contamination. Include proper drainage of spilled water into a soak pit at a safe distance from the well opening;
5. Estimate the maximum yield of wells; over-extraction might bring about saline intrusion (in coastal areas) or cause the well to dry up. If necessary, ration the water supply.

In many emergency situations, water has to be trucked to disaster stricken areas or camps. Water tankers may be obtained locally from commercial water delivery companies, dairies, breweries, bottling plants, etc. All trucks should be inspected to determine fitness, and cleaned and disinfected before transporting water. As a rule, gasoline, chemical, and sewage trucks should not be used.

One of the reasons for recommending a higher residual chlorine level in disaster situations is to provide extra disinfection capacity to control contamination in temporary open storage tanks (primarily inflatable rubber). Risk of contamination of these tanks can be significantly reduced by providing a tap (if possible) or siphon to allow direct withdrawal of the water from near the bottom of the reservoir rather than "dipping" and possibly contaminating the tank. When such a tap or siphon is installed the reservoir can also be covered (e.g., with plastic sheeting). Closed water bladders should be given priority when ordering water reservoirs for emergency situations to circumvent the risk of outside contamination.

If *locally available*, mobile water purification equipment may be used in emergencies. However, such plants require skilled operators, auxiliary power, and maintenance and repair facilities, and they only produce limited amounts of drinking water. Extreme caution should be taken before requesting mobile equipment as part of emergency supplies. Experience shows many failures because imported equipment was not suited for the conditions at the disaster site. Shipment of mobile treatment plants always has low priority because they are expensive, bulky, and occupy valuable space.

Mass Distribution of Disinfectants

The mass distribution of tablets, powder, or liquid disinfectants should only be considered in these conditions:

1. Affected persons have experience in their use;
2. Affected persons can receive training in their use immediately after the event through a vigorous education campaign;
3. Appropriate water storage containers are distributed;
4. Public health or community health workers assist in ensuring the appropriate and continued use of the tablets;

5. A distribution network is in place to ensure a proper and continuous supply as needed throughout the emergency phase and in the early rehabilitation phase.

In general, individuals in small and controlled groups may be given such disinfectants to purify small amounts of drinking water for one or two weeks. Every effort should be made to restore normal chlorination, and to protect individual wells and storage reservoirs. This can be accomplished by sealing cracks in well casings and reservoir roofs, providing adequate drainage around wells, and roofing reservoirs.

FOOD SAFETY

Poor hygiene is the major cause of food-borne illness in disaster situations. Where feeding programs are used, as in shelters or camps, kitchen sanitation is of utmost importance. Utensils must be washed in boiled or treated water, and personal hygiene should be monitored in individuals involved in food preparation.

Food supplies should be stored in containers that will prevent contamination by rodents or insects. Refrigeration may have to be improvised.

BASIC SANITATION AND PERSONAL HYGIENE

Many communicable diseases are spread through fecal contamination of drinking water and food. Therefore, every effort should be made to ensure the sanitary disposal of excreta. Emergency latrines should be made available to the displaced, refugees, relief workers, and residents in areas where toilet facilities have been destroyed. Even if toilets are physically intact, they cannot be flushed without a water supply. Lime should be used in communal trench latrines to reduce the development of methane gas and odors. If no sanitation facilities are available, people should bury their excreta.

Personal hygiene tends to decline after natural disasters, especially in densely populated areas and where there are water shortages. The following measures are recommended:

1. Provide basic hand washing facilities (shelters, temporary settlements and camps);
2. Provide washing, cleaning, and bathing facilities (camps for refugees and displaced persons);
3. Make adequate amounts of water available (disaster stricken areas and camps for refugees and displaced persons);
4. Avoid overcrowding in sleeping quarters;
5. Launch education campaigns on personal hygiene, basic sanitation, and waste management.

Wastewater from camps for refugees and displaced persons, field hospitals, feeding centers, washing facilities, etc., requires proper disposal. The most common means is through a soak away, seepage pit, or absorption trench.

SOLID WASTE MANAGEMENT

Solid waste management often poses a special problem in emergency situations. In the aftermath of disasters authorities not only have to deal with refuse and garbage, but also with debris from buildings, utilities, trees, plants, and dead animals. The rapid commencement of debris removal is very important for the rehabilitation efforts. Clearing roads, for example, not only re-establishes access routes, but has a positive psychological impact on the population.

Sanitary disposal of refuse and other waste is also the most effective way to control vector-borne diseases. Garbage collection should be re-established as soon as possible in stricken areas. Burying or burning organic solid waste is recommended and open dumping should be avoided. Carcasses awaiting burial should be sprinkled with kerosene to protect them from predatory animals. Burning large carcasses is difficult unless special incinerators are built, which require huge amounts of fuel.

Heavy equipment will be necessary for debris removal, solid waste collection, and operation of the disposal site. Pre-emergency arrangements with private equipment owners may facilitate their services. The general public should be advised on sanitary waste handling where no services can be provided (such as burning or burying refuse in yards).

Established disposal sites might be inaccessible or unusable for a prolonged period, and new sites may have to be established. Great care must be taken in selecting these sites, since once disposal commences in an area, it often becomes a permanent dump site. Building debris can be used to improve access roads or in other areas where in-fill is needed. Other bulky materials should be flattened using bulldozers, if available.

Special care must be taken when disposing of hazardous materials (e.g., damaged high voltage transformers containing PCBs). Potentially hazardous waste must be safely stored in a place where it can be retrieved later for proper identification, recovery, treatment, and/or disposal.

VECTOR CONTROL

Control programs for vector-borne diseases should be intensified in the emergency and rehabilitation period, especially in areas where such diseases are known to be endemic. Of special concern in emergency situations are: leptospirosis and rat bite fever (rats), dengue fever and malaria (mosquitoes), typhus (lice, fleas), and plague (fleas). In flooded areas rats will escape their burrows in search for dry hiding places, often in dwellings. Flood waters provide ample breeding opportunities for mosquitoes. Dead animals and other organic waste provide food for rats and other vectors.

The following are essential emergency vector control measures:

1. Resume collection and sanitary disposal of refuse as soon as possible;
2. Conduct public education campaigns to eliminate vector breeding sites in and near the home and on measures to prevent infection, including personal hygiene;

3. Survey camps and densely populated areas to identify potential mosquito, rodent, and other vector breeding sites;
4. Eliminate vector breeding sites permanently by draining and/or filling in pools, ponds, and swamps; overturning or removing receptacles; covering water reservoirs; and carrying out sanitary disposal of refuse;
5. Resume indoor spraying if used earlier as a routine control method in flooded areas;
6. In areas where typhus is known to exist, apply residual insecticide powder to louse-infested persons, their clothing, and bedding in camps and temporary settlements (use DDT or Lindane, or alternatively, Malathion or Carbaryl, depending on local resistant strains);
7. Store food in enclosed and protected areas.

Well-organized control of mosquito breeding sites greatly reduces the need for outdoor spraying, but if surveys show it is needed, local resources should be employed. Consideration should be given to the high cost of outdoor spraying and its limited benefits.

Vector control measures should be associated with other health measures, such as malaria chemoprophylaxis, to reduce or eliminate the risk of infection.

Successfully controlling houseflies and rodents is nearly impossible in the early aftermath of a natural disaster. The only acceptable measures against such pests are environmental sanitation and personal hygiene.

BURIAL OF THE DEAD

The health hazards associated with cadavers are minimal. Especially if death resulted from trauma, corpses are quite unlikely to cause outbreaks of disease such as typhoid fever, cholera, or plague. If human bodies contaminate streams, wells, or other water sources, they may transmit gastroenteritis or food poisoning syndrome to survivors.

Despite the negligible health risks, dead bodies represent a delicate social problem. The normal local method of burial or cremation should be used whenever possible. Burial is simplest and the best method if it is ritually acceptable and physically possible. Cremation is not justified on health grounds and mass cremation requires large amounts of fuel.

Before burial or cremation, bodies must be identified and the identification recorded. In many countries, certification of death or an autopsy must precede the disposal of the body. Incorporating a waiver paragraph into legislation governing disaster situations should be considered.

PUBLIC INFORMATION AND THE MEDIA

Besides specific measures already mentioned, public information should be disseminated about available environmental health services and resources, their location, and which authorities should be notified of specific problems. This helps the public to understand the extent of the emergency, reduces confusion, and improves the effectiveness of emergency environmental health activities.

The media will play an important role in providing such information to the public. It is essential that authorities and media practitioners have a common understanding of the objectives of information distribution as well as their respective roles in the disaster. Pre-emergency meetings or seminars to clarify these roles and responsibilities are strongly recommended.

CHAPTER 9.
FOOD AND NUTRITION

The nutritional status of a population depends on the availability of food, its consumption, and its biological utilization. A natural disaster may affect the nutritional status of the population by affecting one or more components of the food chain depending on the type, duration, and extent of the disaster, as well as the food and nutritional conditions existing in the area before the catastrophe.

Slow-onset disasters such as drought are more likely to affect long-term nutritional status than sudden-onset disasters such as earthquakes and hurricanes. Not all sudden-onset disasters produce food shortages severe enough to cause harmful changes in the nutritional status of the population. The effect of any type of disaster on the nutritional status of the affected population is never immediate. Large-scale food distribution is not always an immediate relief priority, and its long-term implementation may, in fact, produce undesired effects.

To plan and implement successful food relief operations, nutrition workers responsible for humanitarian operations must be familiar with the possible nutritional outcomes of specific types of natural disasters, as well as the food and nutrition situation in the affected area prior to the disaster. A nutrition officer trained in emergency management must be part of the disaster planning and response teams.

The immediate steps for ensuring that a food relief program will be effective include: (1) assessing the food supplies available after the disaster; (2) gauging the nutritional needs of the affected population; (3) calculating daily food rations and needs for large population groups; and (4) monitoring the nutritional status of the affected population.

EXPECTED CONSEQUENCES OF DISASTERS ON THE FOOD CHAIN

Hurricanes, floods, land- or mudslides, volcanic eruptions, and sea surges directly affect food availability. Standing crops may be completely destroyed, and seed stores and family food stocks may be lost, especially if there is no warning period. Volcanic eruptions can cause widespread crop destruction: food crops may be burned, defoliated, and buried under ashfall; reduced photosynthesis resulting from ash clouds limits subsequent production.

Earthquakes, on the other hand, generally have little direct impact on the long-term total availability of food. Standing crops are unaffected, and food stocks can often be salvaged from family, wholesale, and retail stores. However, temporary food problems may result as a consequence of the breakdown of the transportation and marketing systems. If an earthquake strikes during a labor-intensive period such as harvest, the loss of labor from death or its diversion from agriculture may cause short-term scarcities.

The most likely consequence of any kind of sudden-impact disaster will be the disruption of transportation and communications systems and upheavals in routine social and economic activities. Even when food stocks exist, they may be inaccessible due to disruptions in the distribution system or the loss of income with which to buy food. Destruction of cash crops also will have an effect on the economy of families. When destruction of a greater magnitude occurs, leading to the death of livestock and the loss of crops and stored foodstuffs, the short-term dilemma can leave a more severe, long-term crisis in its wake. Moreover, evacuation and resettlement of communities during the post-disaster period are often necessary, creating foci in which total food supplies will have to be provided for the duration of the encampment. Hospitals and other institutions may require emergency food supplies as well. Livestock may have to be sacrificed if they cannot be fed, and they are likely to die when vast tracts of land are flooded for long periods. While the meat can be used immediately for distribution among the affected population, or salted for later distribution, in the long run it results in food and economic shortfalls.

The effect of disasters on the biological utilization of food, that is, intestinal absorption and subsequent utilization of nutrients, is indirect, and dependent on factors such as the impact of the disaster on the environment, particularly on water supply and sanitation. This is an issue of concern, particularly in regard to gastrointestinal infections since they affect the absorption of nutrients. Other infectious diseases increase the demand for nutrients. These effects are more likely to occur among the young and vulnerable groups. If there is an increase in undernutrition rates among young children soon after a disaster, it will most likely be the effect of gastrointestinal illness rather than actual food shortages. This is something to keep in mind in the implementation of surveillance mechanisms. Outbreaks of infectious diseases are uncommon after natural disasters, especially in the Americas.

POSSIBLE ADVERSE EFFECTS OF LARGE-SCALE FOOD DISTRIBUTION

The decision to distribute large amounts of food, although made at the political level, should be based on the most accurate information available. If unnecessarily large quantities of food are brought into an area, this may hinder recovery. Food distribution requires transport and personnel that may be better employed in other ways, and small farmers may face hardship due to depressed market prices. Perhaps the most serious side effect is that maintaining a population by free food distribution, if not accompanied by essentials such as seeds and tools needed to restart the local economy, may create dependence on relief.

SETTING PRIORITIES

The priorities in alleviating food problems are to: (1) supply food immediately where there appears to be an urgent need, namely to isolated populations, institutions, and relief workers; (2) make an initial estimate of likely food needs in the area, so that steps can be taken toward procurement, transport, storage, and distribution; (3) locate or procure stocks of food and assess their fitness for local consumption; and (4) monitor information on food needs so that procurement, distribution, and other programs can be modified as the situation changes.

IMMEDIATE RELIEF

During the first, usually chaotic, days after a disaster strikes, the exact extent of the damage is unknown, communications are difficult, and the number of people affected seems to double by the hour. Food distribution must start as soon as possible to keep people fed, rather than prevent clinical malnutrition. Given the large variety and small stocks of commodities sent in as aid by governments, agencies, private organizations, and individuals, however, food distribution is initially a day-to-day exercise. Planning nutritionally sensible food rations during this period is impossible. What matters during this "chaotic stage" is to provide a minimum of 6.7 to 8.4 Megajoules (1,600 to 2,000 kcal) per day, per person.

As an immediate relief step, available food should be distributed in sufficient quantity to any group that is at high risk or appears to be wanting, to ensure survival for one week (3 or 4 kg per person). Food may be included automatically, for example, in supplies sent to communities isolated by earthquake or displaced by flooding. Where fuel shortages are likely, it may be better to distribute cooked food such as boiled rice or bread rather than dry food.

No detailed calculations need be made of the precise vitamin, mineral, or protein content of the food distributed in the initial phase, but supplies should be acceptable and palatable. The most important thing to be provided is sufficient energy. If no other items can be obtained, distribution of a cereal alone will be sufficient to meet basic nutritional requirements. When a population can find some of its own food, it may be possible to supply only part of the ration, or one food item that complements the basic or staple food lacking in their available supplies.

ESTIMATING FOOD REQUIREMENTS

As soon as possible after a disaster, a rapid assessment of the food and nutrition situation should be made to get a rough estimate of likely bulk food items needed. This is based on the population affected, its composition, distribution (for example, isolated villages, refugee camps), and locally available foods. This will enable managers to take the necessary steps to locate and procure stocks, storage, and transport. Hoarding is not uncommon and leads to over-response.

In the absence of detailed information, an estimate of food requirements must be based to some extent on judgment in the light of the initial assessment, but it should take into account the following factors: (1) the probable effect of the disaster on food availability (e.g., a tsunami may have destroyed all household supplies); (2) the approximate size of the population affected; (3) normal food supply and variations within the area (e.g., the approximate percentages of the population who are subsistence farmers and those who depend wholly on purchased food); and (4) the impact of seasonal factors. In subsistence areas just before the harvest, for instance, household and traders' stocks may be depleted and the population may be more dependent on the market.

The nutrition officer should prepare estimates of foods on the basis of a family unit (usually considered to consist of five people) for one week and one month. Logistically, food distribution on a family basis for one month may be considered the most practical approach. The nutrition officer also should prepare estimates of commodities required by large population groups, for instance, on the basis of 1,000 people for one month. Two simple and useful rules of thumb are: (1) 16 metric tons of food sustain 1,000 people for one month, and (2) to store one metric ton of food, about two cubic meters of space are needed. Proper storage is extremely important to avoid food losses due to rain, pests, or looting.

When calculating the composition of daily rations, the following points should be kept in mind: (1) the ration should be kept as simple as possible; (2) to facilitate storage and distribution, nonperishable food commodities that are not bulky should be chosen; and (3) substitution of items within food groups should be allowed for.

The food ration should be based on three food groups: a staple, preferably a cereal; a concentrated energy source such as a fat; and a concentrated source of protein, such as salted or dried fish or meat. In practice, the diets will be dictated by the availability of ingredients. A standardized ration may be impractical as availability will change daily and according to areas.

Whenever possible, vulnerable groups should receive a food supplement in addition to the basic diet. Among these groups we include children under 5 years old, who are growing very fast and may suffer permanent damage if malnourished, and pregnant and lactating women, who require more nutrients. Breastmilk is the best food for infants under six months of age, and Health Disaster Coordinators should not allow the emergency situation to become an excuse for flooding the country with infant formula.

PROCUREMENT

If the calculated amount of food required exceeds immediate local availability, and if it is anticipated that food will have to be distributed for several months, steps must be taken to obtain food from elsewhere in the country or abroad. A rough estimate of local food transport requirements should also be made for this contingency.

Food for the initial emergency distribution phase should be obtained from national government or wholesaler stocks, or from bilateral or international development agencies (e.g., World Food Program, NGOs).

If large quantities of food are required from abroad, procurement and shipping may require several months. Approaches to suitable agencies should hence be made at the earliest possible date. It is critical that Health Disaster Coordinators advise potential donors of the eating habits and preferences of their populations. Food not eaten is of no nutritional benefit.

The need for special infant foods ("baby foods") immediately after disasters is often exaggerated. Improving maternal nutrition and assisting mothers economically is more cost-effective and safer than airlifting strained baby foods. Since vitamin requirements are of little concern during the acute emergency phase after sudden-impact natural disasters, multivitamin tablets should not be requested as a separate relief item. The population's specific vitamin and mineral needs will have to be assessed for the long-term.

SURVEILLANCE

If long-term food supply problems seem likely, as in areas with subsistence agriculture and poor communications, the nutritional status of the community should be monitored. This can be accomplished by making regular physical measurements of a suitable sample of the population. Since young children are the most sensitive to nutritional changes, the surveillance system should be based on them, remembering that the most serious malnutrition results from an acute exacerbation of chronic under-nutrition. In emergency situations, weight-for-height will provide the best indicator of acute changes in nutritional status. If height and weight cannot be measured, arm circumference, which is simple and easy to measure, may be used to gauge changes in communities.

As the results of the first needs assessments become available, more accurate information will make it possible to adjust preliminary estimates of the proportion of the population most in need of long-term food distribution. Surveys of need should make sure to cover not only food availability, but also identify areas where problems of labor, tools, marketing, and other variables affecting distribution have arisen. As soon as an area is able to return to normal consumption patterns, distribution should be phased out.

Chapter 10.
Planning, Layout, and Management of Temporary Settlements and Camps

Health authorities will not usually be directly responsible for setting up and managing camps and temporary settlements. Since many aspects of camp management affect the health of the occupants, however, the Health Disaster Coordinator should be involved in decision making as early as possible.

PLANNING SETTLEMENTS AND CAMPS

Temporary settlements or more permanent camps arise in a number of ways. After floods, people can be forced to move to higher dry ground. Such settlements often disband spontaneously when flood waters recede, but may become long-standing if a flood seriously damages agricultural or building land.

After earthquakes or destructive winds, some people who have lost their own houses may be unable to find lodging with relatives and friends. When after-shocks occur and a continuing risk is perceived, people often move into open spaces, parks, and fields.

Humanitarian assistance should be provided to people in or close to their homes. Whenever possible, the deliberate creation of camps should be avoided. More problems will generally be created than are solved because camps and temporary settlements present the greatest potential for communicable diseases once the immediate disaster has passed, and often become permanent even when that is not intended.

It might be expected that providing services to a camp would encourage the population to stay and become dependent on relief. Though this may become true after long periods, it is unusual in the short term. People generally prefer to return to their normal lives and surroundings, and when they become dependent on relief it is usually for want of an alternative.

SETTING UP CAMPS AND SETTLEMENTS

There are two objectives in setting up camps and settlements. The first is to ensure a standard of living for the inhabitants as close as possible to that among similar groups in the country outside the camps. Particularly in temporary settlements, voluntary workers and agencies sometimes provide much better services, food, and housing than the occupants had before and will have after the emergency. This causes friction with the surrounding population, and gives refugees expectations that the national authorities cannot possibly meet. The second objective is to minimize both capital and recurring costs and the degree to which continuing external administration is required in running a camp.

SITE SELECTION

Locations for camps and settlements should be defined in disaster plans. If this has not been done, a suitable location should be chosen as soon as possible, since that will influence all other decisions about layout and provision of services. The site should be well drained, not prone to seasonal flooding, landslides, tidal waves, or sea surges, and located as close as possible to a main road to ease supply problems. If international support is anticipated, a site with reasonable access to an airport or port should be selected. Locating a camp away from existing urban areas makes access easier and can minimize administrative problems, but for long-term resettlement, a site close to an existing community facilitates the provision of transport and employment.

Around urban areas, where the pressure on land is high, camp land may be available precisely because it is unsuited for residential use. The possibility of acquiring land by purchase or from government holdings should be considered.

CAMP LAYOUT

Permanent communities are characterized not only by their buildings and streets but also by their social cohesion. Since people share services and have common needs, mutual obligation systems evolve that regulate behavior in regard to property protection, waste and water disposal, latrine use, and play areas for children. In shantytowns these mechanisms may be inadequate, but in camps they will be lacking entirely. Such lack of social cohesion contributes to the spread of disease (e.g., by failure to use latrines) and makes camp administration more difficult. Adequate and early attention to physical layout will minimize such problems.

Camps should be laid out so that a small cluster of families is grouped around communal services. Access to a set of services (latrines, a water point) should be limited to a fixed group of people, and individual "communities" within the camp should be small enough to encourage the development of social structures. Many administrative tasks such as latrine maintenance and disease surveillance can be partially delegated to these groups instead of being assigned to an employed workforce. The camp can be expanded with no reduction in the quality of services by adding units at the periphery. Areas must be set aside for administration, reception and administration of camp residents, warehouse facilities, supply distribution sites, and recreation areas.

Grid layouts with square or rectangular housing areas intersected by parallel roads, which were widely used in the past, have the advantages that water, drainage, and power systems can be incorporated easily into the camp plan where land is limited, and they can accommodate high density population. This may also be disadvantageous since it is likely to encourage the spread of disease. Grid camps are relatively unsuitable for family occupation and should be avoided, particularly for long-term use.

CAMP SERVICES

Water Supply

Proximity to safe water sources is one of the most important criteria for site selection. If the camp is close to a public water supply, a connection may be possible and an important problem solved. Other systems and sources such as self-contained pumps or purifiers may be used, but they are more costly and require regular maintenance. In some areas, tube or dug wells may provide cheap drinking water of high quality.

Contamination of water in temporary storage facilities such as collapsible tanks or household containers is common. Adequate water chlorination and daily residual chlorine and bacteriologic monitoring will prevent illness. The UNHCR recommends a minimum of 15 liters of clean water, per person, per day for domestic needs.

Excreta Disposal

Adequate sanitation is an essential component of diarrheal disease prevention. At least one latrine should be provided for every 20 people, and latrines should be sited for easy access from any part of the camp to encourage their use. Ideally there should be one latrine per family.

Health Services

If the camp is well organized and has adequate sanitation, water, and food supply standards, health conditions will be similar to those in the general population. Unless there is clear medical justification, providing a higher standard of care to camp residents than to the general population should be avoided. Health services can be provided by assigning volunteers or government health workers to the camp or enlarging the capacity of the nearest fixed health facility. The focus of the health services should be mainly on prevention of specific communicable diseases and the establishment of a health information system.

CHAPTER 11.
COMMUNICATIONS AND TRANSPORT

Effective management of health relief requires access to and control of adequate transport and communication. Because the health sector's resources are usually insufficient to meet those needs, advance planning is particularly important to ensure that other institutions and sectors provide sufficient support in the event of a disaster. As part of pre-disaster planning, the Health Disaster Coordinator should make arrangements with entities such as the ministries of transport, public works, and communications; the armed forces; nongovernmental organizations; private passenger and freight transport companies; private and state telecommunications companies; and ham radio operator clubs.

Responsibility and coordination for emergency government transport and communications should be centralized in a single office of the National Emergency Committee, which can coordinate their use for defined relief needs. The importance of developing a good working relationship with national telecommunications agencies and/or private sector telecommunications service providers cannot be overemphasized.

TELECOMMUNICATION

Adequate emergency communication facilities are essential for maintaining rapid and reliable contact with health facilities and relief personnel in the field, as well as with governmental, nongovernmental, private, and international agencies involved in the relief effort.

In most countries, the government allots specific radio frequencies and equipment to the military, fire and emergency services, police, ham radio operators, the private sector, and others in accordance with standards established by the International Telecommunication Union (ITU), a United Nations agency.

Telecommunication technologies and services have undergone tremendous growth during the past decade. Voice messaging, cellular and satellite telephone, and teleconferencing are among the options offered by an increasing array of service providers. The administration of telecommunications assets and services also has changed. State control has been ceded to private companies in some countries; in other cases, State and private companies cooperate in managing the system.

Reliable, low-cost service is still not available everywhere in the world, however. The increased recognition of how vital communications services are, both nationally and internationally—which has been fueled by new portable satellite services and the enormous popularity of the Internet—offers the hope of expanded accessibility of telecommunications. Following is a brief overview of available services and their reliability and usefulness to disaster managers.

Radio Communication

Radio systems offer many advantages in disaster situations. However, while operating costs are low, the costs of installing and maintaining an efficient system can be high. In most countries, fire and emergency medical systems, military, police, and related institutions maintain some type of radio link, although each service will tend to operate its system independently of others. The health sector should ensure its connection to the national system or systems, and take advantage of the technical expertise offered by capable personnel. There is a wide range of available technology and possible uses, which are summarized below.

High Frequency (HF) Single Sideband (SSB) Radio

Disaster managers in the field most frequently use HF radio to communicate over long distances. This communication is point-to-point and permits voice and low-speed data communications between and among fixed installations at field headquarters and regional offices. Mobile HF SSB units can be used in a similar manner (although by definition, "mobile" units are considered to be permanent installations in vehicles), as can transportable units, which are integrated communications packages designed to be deployed at single locations upon short notice.

A significant advantage of HF SSB networks is that hardware costs are minimal (US$ 4,000–US$ 5,000 for the basic components of a voice-and-data system), and use is free. A disadvantage is that, because of its wide use, it is difficult to get allocation of the dedicated HF frequencies required to operate the system. HF transmissions are also subject to propagation effects that occur daily and seasonally.

Effective distance of HF voice communications ranges from 2,000–3,000 km to 10,000 km, which is usually sufficient for communications between field operations and national headquarters. The use of advanced technology (e.g., the Pactor Level 2, Clover, and other data modes) along with the use of enhanced modems permit effective data communications worldwide.

Very High Frequency (VHF) Hand-Held Radio Communication

For short distance communication (within cities and within geographic regions of approximately 100 km) use of VHF hand-held radios is ubiquitous among national authorities, United Nations agencies, and NGOs for communication among staff. As are HF radios used for longer distance, VHF radios are relatively inexpensive to purchase and free to operate. However, the use of VHF equipment is subject to the delivery of a license with a limited number of assigned frequencies, a process requiring a significant amount of negotiation with local telecommunications authorities. In the absence of regular telephone communication, VHF radios provide a basic and vital administrative function. Another important function is security, as it is possible to maintain contact with staff traveling from one part of a city to another.

Amateur Radio Operators

Amateur, or "ham" radio operators have historically been the first to establish and operate communications networks locally for governmental and emergency officials during and immediately following a disaster. Amateur radio facilities can generally be characterized as having a high survival capability. Although amateur radio operators are most likely to be active after disasters that cause power outages and destruction of telephone lines, they frequently support the delivery and relay of pre-disaster and warning information. Amateur radio operators are generally well-motivated, willing, and prepared to work under extreme conditions encountered during acute emergencies, where both solid technical knowledge and the ability to improvise are required. Although most operators belong to organized groups and show a great sense of discipline and responsibility, the accuracy of their reports may vary widely. Direct, close coordination of these groups by emergency telecommunications managers is critical to avoid the danger of transmitting inaccurate, unconfirmed, or unreliable information.

Governments license amateur radio operators in most countries. Some governments severely restrict the use of amateur radio operators. The International Amateur Radio Union (IARU) coordinates the activities of amateur services and actively supports their introduction in those countries where their value has not yet been fully recognized.

Radio Paging Service

Radio paging is increasingly common in most countries. Its coverage can range from local to international and it is of unquestionable value to disaster managers. Access to and reliability of these systems during disasters depends on a variety of factors ranging from the availability of telephone, cellular, and/or satellite lines to interconnect and operate the system, and the availability of independent sources of electrical power, to the quality of the actual paging transmitter.

While the majority of paging systems are one-way and do not guarantee delivery to the recipient, 1½-way (message receipt acknowledgment) and two-way systems are emerging, often with electronic mail links.

In countries where there is sizable GSM (mobile telephone) penetration, traditional paging systems have been supplanted by the SMS (short message service) which is built into GSM protocols.

Although limited resources may not make it possible to routinely use pagers, if paging service is available, it may be advisable to lease the service for essential personnel in disaster situations. It must be kept in mind that most radio paging services rely on terrestrial infrastructure which is vulnerable during disasters. This is the case to a lesser extent with satellite-based paging systems.

Terrestrial Telecommunications

Traditional terrestrial telecommunications services, the most characteristic of which is telephone service provided via telephone wires, have been costly to install, difficult to repair, and vulnerable to disasters, particularly in remote areas of developing countries. If even one telephone pole is incapacitated in a terrestrial network, all communications past that point are affected until that pole and its connection to the system can be repaired. Thus, although they may play a role in disaster planning and early phases of warning, terrestrial communications cannot be relied on for continuous use during the disaster onset and in the acute phase of disaster response.

Even if the telephone system is not damaged by the disaster event, it is likely to be made unreliable or unusable because of heavy demand by the affected population. Outages due to overloaded circuits can last anywhere from several hours to several weeks. Dial tones, too, can be affected by power outages and overloads and can be a further obstacle to disaster management.

Health sector disaster managers should develop and maintain good relationships with local and national telecommunication service providers and work with them to develop disaster services and emergency protocols based on the infrastructure at hand.[1] National governments should be encouraged to strengthen their terrestrial telecommunications infrastructure and to make it resistant to the type of hazards present in their countries.

Satellite Communications

Fixed Satellite Service

Early satellite communications service and infrastructure were developed first within large urban areas. Then, through rapid advancements in space technology, population centers were connected. Early communication satellites developed in response to the demand for their services.

The phenomenon of television resulted in increased requirements for greater satellite capacity which was, again, dedicated largely to population centers. Because the technology of these early fixed satellite services did not provide for powerful transmitters, large, complex, and expensive earth stations had to be used to receive and send signals from and to the satellites. They were used as regional or national gateways for major telecommunications trunking services and for television distribution. They are still, for the most part, limited to communications within and between capital cites and large urban areas.

Space segment providers and satellite designers soon recognized the need to reduce the size and expense of the ground hardware and launched a new generation of services that relied on more versatile and powerful satellites. Because these new satellites transmitted more powerful signals, the new ground hardware size and power requirements were significantly reduced.

The capital cost for hardware and the recurrent cost of satellite time were also reduced. In practical terms, this meant a move from stationary earth stations to transportable (*not* portable) systems. These changes in service made possible the advent of very small aperture terminals, or VSATs. The applicability of VSAT services could include linking, in a permanent or semi-permanent network, national health sector disaster managers. This is still costly and it must be remembered that, as with terrestrial telecommunications links, fixed satellite service infrastructure is

---

[1]See Mark Wood, *Disaster Communications Manual*, available in print from The Disaster Relief Communications Foundation or on the Web at http://www.reliefweb.int/library/dc1/dcc1.html.

susceptible to damage or destruction at the onset of the disaster. Unlike its terrestrial counterparts, however, if one link goes down, all others are not affected.

Global Mobile Personal Communications by Satellite (GMPCS)

In the near future there will likely be dozens of low earth orbit (LEO) satellite systems covering the entire world in addition to the existing geosynchronous satellite networks. They will consist of from 1 to as many as 325 satellites per system and will be a part of a new category of service, GMPCS. These systems offer disaster managers the promise of easy-to-use, reliable, and affordable communications, regardless of the nature of a disaster, its location, or the terrain of the affected area.

These systems will have a wide range of capabilities ranging from narrow band (data only) to broad band (allowing video, voice, and data) communications. GMPCS technologies will likely be of such low cost as to be affordable to all sectors and, thus, should be of great interest to health sector managers.

Despite their promise, in the near-term disaster managers should be cautious when deciding what new and emerging technologies to use and invest in. It is advisable to explore a mix of technologies, including, but not limited to, new GMPCS services until such time as their strengths and weaknesses have been proven.

Mobile Satellite Service

Mobile satellite services were first developed for maritime applications and now are broadly used for aeronautical and land-based purposes. Mobile satellite services are less expensive than traditional fixed satellite services. They are easily transportable and are not dependent on the terrestrial telecommunications infrastructure. They are far less vulnerable to natural disasters and, because they can be used reliably to send data or call anywhere in the world, their use in the field has grown rapidly.

Although lower in cost, they are not inexpensive, and are still used almost exclusively by United Nations agencies and larger NGOs. Although some national/domestic systems are available, the most widely used is that developed by INMARSAT, an international consortium. Costs range from US$ 4,000–US$ 35,000 for hardware and US$ 1–US$ 13 per minute for use.

Additional systems at either end of the technology and cost spectrum are rapidly becoming available. These range from Iridium, featuring hand-held telephones that allow for voice calling from anywhere in the world, to Orbcomm, which offers worldwide low data messaging and data collection from fixed sites or hand-held units.

Electronic Mail Services

Communication by electronic mail has undergone explosive growth. Initially, it was possible to communicate within closed networks, but with the opening of the Internet to the civilian world, millions of individuals and institutions can exchange information from almost any point on the planet. Electronic mail requires a computer with a modem, access to telephone or other telecommunications service, an Internet account, and some training.

The Internet is very useful not only in the immediate post-disaster phase, but also in prevention, preparedness, and mitigation aspects of disaster management.

It is inexpensive to use compared to traditional communication systems. The service makes it possible to contribute to and use information on Web sites about disasters; form and participate in discussion groups and "virtual" conferences among institutions worldwide; and send documents and graphics. It allows free exchange of information among interested parties, with a minimum of bureaucratic restrictions.[2]

It should be understood, however, that most Internet Service Providers rely on terrestrial telecommunications infrastructure, making Internet service vulnerable during disasters. This is an issue not only in terms of the Internet's reliability as a communications tool, but also because Internet users have become accustomed to "storing" valuable data on the servers of their Internet Service Providers and the safety of this data may be in jeopardy during disasters.

Teletype

Teletype is a well-known system, but it has been largely superseded by other systems in emergency communications because of high cost and slow transmission speed. It is still used in certain areas, such as in banking, since it offers security in certain types of data transmission. Availability and access are limited, and its use in emergency situations should not be considered.

Donated Radio Communication Equipment

After a major disaster, there may be an outpouring of offers of donations from countries, organizations, and businesses. Supplemental radio equipment is sometimes included in these offers, but often the radio units are delivered well after they are needed. To make sure that the equipment is of use in the disaster, the donor should be informed of certain technical characteristics, such as: transmitting and receiving frequencies; output requirements (wattage); the country's voltage and amperage; number and type of units needed (hand-held, mobile, base stations); type and quantity of antennas; necessary material for installation (coaxial cable, tools); and the need for specialized personnel.

It is important to keep in mind that radio communications are affected by many factors, and therefore are never completely reliable. Geographical features, atmospheric conditions, urban density, electromagnetic radiation from power transmitters, condition and quality of antennas and transmission cables, quality and capacity of equipment, as well as sunspots and solar eruptions, can all have a negative effect on communication quality. All communication systems are dependent at some stage on radio: telephones have microwave links between central exchanges, and links with satellite; the Internet depends on links between satellites and earth stations; radio pagers depend on radio signals between the central exchange and the beeper. Disaster managers need to be aware of potential failures in these systems at the disaster planning stage. They must also have an inventory of equipment available.

---

[2]Recommendations made at the Meeting on Health Crises and the Internet, held in Bogotá, Colombia, November 1997, can be viewed on the Internet (http://www.paho.org/english/ped/ped-internet.htm).

Effective communication after disaster does not depend only on the nature and quantity of equipment available. The willingness of authorities to exchange and communicate specific and detailed information to the public, other governmental agencies, and the international community is of primary importance.

TRANSPORTATION

As in the area of communications, the health sector should coordinate with national institutions for logistic support in transportation. It is crucial to identify the entity responsible for coordinating transport in emergency situations. Arrangements should be made during the disaster planning phase with the following: ministry of public works or transport, the armed forces, the police, public and private passenger and freight transport companies, shipping companies, airlines, and NGOs involved in disaster relief.

Frequently, these institutions or agencies provide vehicles in case of disaster, and the requesting institution covers the cost of fuel and the salary of operators. Airlines generally transport humanitarian relief supplies at a reduced cost. As part of disaster planning, the health sector should establish relations with selected entities, and identify necessary financial resources for operations.

Inventory of Resources

As part of pre-disaster planning, an inventory should be made of vehicles in the country or province that can be commandeered for relief purposes, and the institution providing them (e.g., ministry of health, social security, municipal health service, NGO, etc.). The inventory should specify the type of vehicles (with emphasis on collective transport, four-wheel-drive cars or trucks, and refrigerated vehicles); their maintenance and fuel requirements; their ability to transport personnel or cargo; location; and the names and contact numbers of persons responsible for authorizing use. The same inventory should include agreements made with public and private transport companies during the disaster planning phases, and names of contact persons. It is important to note that contracting vehicles locally must be done immediately in the event of a disaster, as competition among agencies for vehicles in good condition will be intense.

Transport Equipment Needs

Initial requirements for transportation usually focus on life-saving operations; transporting essential staff, equipment, and patients; conveying specialized personnel to assess and evaluate health status in the affected zones; delivery of necessary health supplies to treatment centers; removal of bodies and animal carcasses; clearing access routes to hospitals and health centers; and transporting international donor representatives and media personnel to and from the affected area.

The need for ambulances is often exaggerated. In the early life-saving phase, demand is very high and almost any vehicle is used. Multi-purpose vehicles where stretchers can be fitted are now routinely used by many organizations. The transport of potable water and fuel are areas of special importance to continued humanitarian health operations.

Equipment Sources

Figure 11.1 shows possible uses and sources of vehicles and logistical support equipment that may be available for use in the period immediately after a disaster. The uses suggested depend on the actual situation (as with the use of boats in a flood), and potential sources will vary from country to country. It is often more realistic to rely on national and local sources rather than on international sources.

The Health Disaster Coordinator should give forethought to the logistical support required to carry out a relief operation. This includes fuels and lubricants, road-clearing and cargo-handling equipment, trained drivers, and vehicle mechanics. Mechanics are needed to ensure that the vehicles in emergency reserve are kept in operating condition.

FIGURE 11.1. Possible uses and sources of transport equipment available during immediate post-disaster period.

Vehicle type	Survey of disaster	Transport health personnel	Evacuation	Supplies to staging area	Supplies to/within affected area	Removal bodies	Removal/burial animal carcasses	Transport media personnel	Road clearing	Water supplies	Fuel supplies	Supply handling	Epidemiological surveillance	Functional ministries	Military	Local commercial dealers	Private/commercial ownership	Foreign agencies/governments	Red Cross other NGO's
Ambulance		✔	✔											✔	✔			✔	✔
Utility 4-wheel drive	✔	✔	✔		✔	✔		✔		✔	✔		✔	✔	✔	✔	✔	✔	✔
Pickup truck		✔	✔		✔	✔	✔	✔		✔	✔		✔	✔	✔	✔	✔	✔	✔
Van/station wagon		✔	✔		✔	✔		✔					✔	✔	✔	✔	✔		✔
Motorcycles	✔	✔						✔					✔	✔	✔	✔	✔		✔
Medium/heavy trucks				✔	✔		✔			✔	✔			✔	✔	✔	✔		✔
Bicycles		✔						✔					✔	✔		✔	✔		✔
Water tankers				✔	✔					✔				✔	✔		✔		
Boats (river)	✔	✔	✔	✔	✔	✔		✔					✔	✔	✔	✔	✔		✔
Barges (+ tugs)				✔												✔	✔		
Amphibious vehicles		✔	✔		✔	✔		✔							✔				
Helicopter	✔	✔	✔		✔			✔							✔			✔	
STOL aircraft	✔	✔	✔	✔	✔			✔							✔		✔	✔	
Cargo aircraft			✔					✔							✔		✔	✔	
Amphibious aircraft	✔	✔	✔		✔			✔					✔		✔			✔	
Fuel tankers				✔	✔									✔	✔				
Bulldozers							✔		✔					✔	✔	✔	✔		
Auto cranes												✔			✔		✔		
Animal transport					✔	✔	✔	✔					✔			✔	✔		✔

Chapter 12.
Managing Humanitarian Relief Supplies

The type and quantity of humanitarian relief supplies are usually determined by two main factors: (1) the type of disaster, since distinct events have different effects on the population; and (2) the type and quantity of supplies available in national inventories prior to the occurrence of a disaster.

Immediately following a disaster, the most critical health supplies are those needed for treating casualties and preventing the spread of communicable diseases. Following the initial emergency phase, needed supplies will include sanitary engineering equipment, food, shelter, and construction material.

There are always delays in the arrival of assistance from abroad. Immediate needs must be met primarily with locally available resources from the affected country and resources from provinces or departments adjacent to the area of impact. Humanitarian relief supplies that must come from neighboring countries or from abroad should be strictly limited to those items that meet specific needs that cannot be supplied locally.

The first humanitarian assistance shipments will arrive at a country's main entry points (airports, seaports, or land border crossings) within 24 to 72 hours of the event, but unloading, sorting, storage, and distribution of supplies will take much longer. The majority of relief supplies usually arrive after the most urgent health needs have already been met with local means.

The main problem in all but the least economically developed countries is not the acquisition of large quantities of new supplies in the event of a disaster, but rather the taking advantage of locally available resources. Identification, sorting, classification, inventory, storage, transport, and distribution of items, especially of unsolicited donations, pose another major challenge.

BASIC PRINCIPLES

The following basic principles regarding humanitarian relief supplies should be kept in mind:

- Sources for emergency supplies should be identified as part of the disaster preparedness process. Ideally, there should be a national inventory of resources that can be used in the event of a disaster.
- Stockpiling supplies exclusively for disaster situations is not recommended, because of the high costs entailed for developing countries. Such stockpiles require very efficient supply rotation systems that are costly to implement.

- When a disaster occurs, rapid damage assessment must be carried out in order to identify needs and resources.
- If external assistance is necessary, requests should be strictly limited to resources not available in the affected zone.
- Disaster managers must be prepared to receive large quantities of unsolicited donations from other areas of their own country, neighboring nations, and the international community. The quality and usefulness of such donations is often questionable.
- When requesting supplies, the time required for shipment and distribution must be considered, and there must be planning for needs that will remain unmet after supplies arrive.
- No supplies or other forms of relief should be sent without first verifying the need for such assistance. It is essential to assign priorities for each container shipped.

THE LOGISTICAL SUPPLY CHAIN

There are four principal components in managing humanitarian supplies:

- **Acquisition of supplies.** This requires determining what items are necessary, how to acquire them, and how to use them to meet identified needs.
- **Transport.** This entails an accurate assessment of readily available and alternative means of transportation to promptly and safely deliver supplies.
- **Storage.** An organized storage system safeguards supplies until they can be delivered to their final destination. The system also assists in anticipating amounts of supplies in reserve for later needs.
- **Distribution.** The ultimate objective of the logistic supply chain is to deliver assistance to the persons affected by the disaster or to the organizations in charge of their use. Balanced and controlled distribution must be ensured to avoid abuses, waste, or damage to the supplies.

These components are linked and complementary, and require very careful coordination to ensure that there are no interruptions in the logistics chain.

SUPPLY MANAGEMENT

The main objective of a humanitarian assistance management system is to strengthen national capacity so that supplies are effectively managed from the moment donors offer assistance and through their arrival and distribution in the disaster affected area.

As in other aspects of disaster management, managing humanitarian assistance supplies cannot be improvised. A standing committee or permanent official should be designated to:

- Establish national policy regarding donations and receipt of emergency supplies;

- Promote transparency by openly circulating information among agencies; and
- Provide cross-sectoral training in management of humanitarian assistance.

The agencies most involved in this field, in addition to the health sector, include civil protection agencies, customs agencies, Red Cross societies, and other NGOs capable of mobilizing national and international assistance.

Information management and coordination of logistic operations are the responsibilities of a variety of sectors outside of the health sector. The health sector should concentrate on management of medical and public health supplies.

Humanitarian relief supplies that arrive following a major disaster, whether natural or complex in origin, cause serious logistical and administration problems for national and international authorities. This is particularly true when the supplies have not been requested and their value in terms of meeting real needs is questionable.

A supply management system should be oriented to resolve the following issues:

- Space and transport are scarce or are not immediately available;
- Time is short—an emergency demands rapid and effective supply distribution;
- Donors and the news media receive a negative impression if local officials are unable to absorb supplies quickly and effectively and at the same time there are urgent appeals for assistance;
- There is no follow-up on offers made by donors;
- One agency may receive an excess of supplies, while another does not receive any;
- Key health personnel lose precious time sorting through donated medications that are of limited benefit.

Supply Management Systems

Long experience in international relief operations has demonstrated the value of a uniform system of supply management. Many recipient and donor governments and organizations involved in the management of supplies use the SUMA system. This system, created by the Pan American Health Organization, was adopted by the World Health Organization as a standard for the general management of emergency supplies (see Annex II).

The main tasks of a supply management system (such as SUMA) include:

- Sorting and identifying humanitarian assistance supplies;
- Rapidly identifying and establishing priorities for the distribution of supplies urgently needed by the disaster-affected population;
- Maintaining inventory and distribution control in warehouses;
- Entering all incoming supplies in a database (national authorities use reports generated from the database for decision-making);
- Registering consignments that are delivered to consignees;
- Keeping disaster managers informed about items available for distribution;
- Keeping national authorities and donors informed about items received.

Sorting Supplies by Priority

Certain incoming supplies satisfy urgent needs, others will be useful at a later phase of the emergency, and some are not useful at all. Supplies are therefore separated according to their priority levels. Incoming packages should be clearly labelled to show their level of priority; this is a key factor in managing storage, transport, and distribution of items.

Classifying Supplies

Incoming supplies are generally mixed together: one shipment may contain articles ranging from medicines to construction materials. Life-saving supplies compete for attention with less important or unsolicited items.

SUMA employs a classification system for humanitarian assistance supplies based on 10 basic categories, which are further divided into subcategories. These 10 categories are:

Medicines;
Water and environmental health;
Health supplies/Kits;
Food;
Shelter/Electrical/Construction;
Logistics/Administration;
Personal needs/Education;
Human resources;
Agriculture/Livestock; and
Unclassified.

Supply Inventory

Immediately after classifying supplies, information about technical characteristics, potency, presentation, packaging units, total quantity, etc., is entered into a database. This information, consolidated into a single list, allows disaster managers to make decisions regarding distribution. An inventory is only useful if it is kept up-to-date. Reports that accurately reflect the availability of supplies are very valuable in the decision-making process. Mechanisms to ensure the accurate generation of such reports are an essential component of any supply management system.

Distribution and Storage of Supplies

Once items have been sorted, classified, and inventoried at the arrival point, they are either delivered to consignees where their receipt is registered, or they are sent to an existing or temporary storage facility. All information regarding distribution is transferred to the Emergency Operations Center.

Follow-up on Donor Offers

Donor countries or agencies may not make immediate shipments to a disaster site, but depending on their capabilities, will offer to send specific materials or equipment. These opportunities are frequently lost if a system is not in place to organize corresponding information. The system should have reliable mechanisms to follow-up on the offers, as well as to keep track of donations received.

Use of Local Stocks

Delays in delivery and high costs of air freight, in addition to lengthy customs procedures (despite international agreements on the free movement of humanitarian supplies), are additional factors to be considered before making urgent appeals for assistance.

Expenditures of air freight are generally subtracted from the total amount a donor allocates for a given disaster. This is another reason for directing efforts to improve the availability and access to resources existing in a country before a disaster occurs.

Every country has normal operating stocks of health supplies found in warehouses belonging to the national and municipal health services, pharmaceutical companies, private health services, nongovernmental organizations (e.g., Red Cross societies, Médecins sans Frontières, CARE, WorldVision), or military and police health systems. These supplies, available in and outside of the affected area, are usually sufficient to meet basic immediate needs during the emergency phase. Even when warehouses are damaged, some stock may be salvageable.

Localized shortages arise in the emergency period because of three main factors: (1) sudden disruptions in normal supply channels and in the availability of supplies; (2) difficulties in locating, accessing, sorting, classifying, inventorying, transporting, and distributing supplies in a disaster zone; and (3) the high consumption of items such as x-ray film and developing chemicals, casting plaster, and dressings.

The effective mobilization and use of available supplies requires that the institution responsible for national disaster management maintain an accurate inventory of resources that could be used in the event of a disaster. The inventory should contain information on the location and condition of the stock. It also should include information on standing agreements for expediting the procurement process of supplies and on mechanisms for transferring supplies to sites where they are needed.

Expired Drugs and Perishable Products

Drugs close to or past their stated expiration dates are often donated or offered. Expiration dates are very conservatively set for some drugs, and with suitable storage, the drugs remain safe and potent for much longer. When such drugs are of particular value, health authorities should decide whether reference laboratory testing and recertification should be arranged. There may be a negative public reaction nationally or internationally if this course is pursued.

Whole blood is often donated from abroad, although its medical need is generally limited. International donors of blood should verify that a need exists, that it cannot be met locally, that the blood can be properly handled, and that its safety can be ensured. It is more suitable to request or donate equipment to collect blood, or a suitable supply of blood substitutes.

ESSENTIAL DRUGS

In 1977, an expert committee of the World Health Organization concluded that "for the optimal use of limited financial resources, the available drugs must be restricted to those called essential drugs, indicating that they are of utmost importance, and are basic, indispensable, and necessary for the health needs of the population."[1] This approach, recommended for non-emergency situations, is critical for health management in emergencies.

The United Nations Children's Fund (UNICEF), the International Federation of Red Cross and Red Crescent Societies (IFRC), the United Nations High Commissioner for Refugees (UNHCR), WHO, Médecins sans Frontières (MSF), and other nongovernmental and governmental organizations have developed lists of essential drugs for disaster relief. They also have established lists of medicines, or "kits," for specific situations, including mental health, surgery, chronic diseases, tuberculosis, cholera, trauma, etc. The lists are designed to meet the general needs of 10,000 persons in specific conditions assuming that there are no other locally available resources. The WHO "kit" has characteristics that can be applied in most situations, and provides a good model for establishing lists of essential needs.[2] These kits are a good source of information for medicines and supplies needed for displaced populations. Although each list is designed to meet the needs and relief functions of an organization in specific situations, they have characteristics in common, so they serve as good models for most scenarios.

Health authorities in each country should prepare in advance their own list of basic medical supplies to be made available immediately after a disaster through locally available stocks. To gain maximum recognition and credibility, a variety of health-related institutions should be involved in developing the lists, including the ministry of health, social security, public and private health services, medical schools, Red Cross societies, pharmaceutical companies, etc.

REQUESTING INTERNATIONAL ASSISTANCE

To maximize the benefit of scarce international assistance to the disaster-affected country, the following guidelines should be followed:

1. A single government official should be designated for channeling emergency international appeals, since otherwise duplication, confusion, and shortcomings will result. In many countries in the Region of the Americas, the ministries of foreign affairs have designated a focal point for emergency situations who is involved in making international appeals for assistance.
2. Potential donors should be asked to provide large amounts of a few items since this simplifies and expedites procurement and shipping.

---

[1]World Health Organization, *The Selection of Essential Drugs; First Report of the WHO Expert Committee* (Geneva, WHO Technical Report Series No. 615; 1977, p. 9). Since 1977 WHO has updated the list of essential drugs every two years. The most recent list is: *Use of Essential Drugs: (Tenth Model List). Seventh Report of the WHO Expert Committee* (Technical Report Series No. 882), Geneva: 1997.

[2]World Health Organization, *The New Emergency Health Kit: Lists of Drugs and Medical Supplies for 10,000 People for Approximately Three Months* (Geneva, WHO/DAP/90.1, 1990).

3. The request should clearly indicate the order of priority, amounts, and formulation (e.g., tablets or syrup). Vague requests for "antidiarrheal drugs," "antibiotics," or "vaccines" must be avoided. The amounts requested should be compatible with the size of the affected population and the anticipated occurrence of trauma and disease. Requests that international donors have considered out of proportion to the magnitude of the disaster have proved counterproductive.

4. Requests should be limited to drugs of proven therapeutic value and reasonable cost. Emergency situations do not justify requests for expensive and sophisticated drugs (especially antibiotics) and equipment which the country could not afford before the disaster.

5. Perishable products and vaccines should not be requested unless refrigeration facilities are available and special handling arrangements can be made at the airport.

6. Supplies will be duplicated if the same list is sent to several donors. Some items may be shipped by a number of suppliers and others not at all. PAHO/WHO can help the country to assess its needs more precisely and inform donors of the most appropriate assistance. The failure to coordinate donations nationally has resulted in time-consuming direct consultations between governments willing to assist, relief agencies, the United Nations Office for the Coordination of Humanitarian Affairs (OCHA), and PAHO/WHO to determine their respective courses of action.

Some donor countries and agencies are reluctant to replace local medical stocks that have been used for emergency purposes and instead want to supply emergency needs directly. This problem is lessened if donors are informed that the depletion of local stocks because of the emergency will restrict rehabilitation of normal medical services. Donors should also realize that their consignments of supplies often cannot be received and distributed in time to be used in treating casualties.

CHAPTER 13.
INTERNATIONAL HUMANITARIAN ASSISTANCE

All countries must strive toward national self-reliance in disaster relief. However, whatever a given country's size and level of development, there are instances when international assistance is needed to provide resources or skills that are not available locally. When disaster strikes, many agencies, associations, groups, and governments offer humanitarian assistance to countries affected by natural disasters. Each has different objectives, expertise, and resources to offer, and several hundred may become involved in any single major disaster. If properly coordinated, international humanitarian assistance is beneficial to disaster victims; if uncoordinated, the resulting chaos and confusion will cause a "second disaster."

Governments must be prepared in advance to assume responsibility for the *coordination* of humanitarian assistance, as this task can hardly be improvised effectively after a disaster. *Operational control* or monopoly by civilian or military institutions is no longer feasible, acceptable, or in the interest of victims. One essential step is to designate a senior health official—the Health Disaster Coordinator—to serve as the focal point for emergency preparedness prior to the disaster, and as the coordinator of health-related humanitarian activities in the aftermath of the disaster. Dissociating the functions of preparedness from those of response is counterproductive, as they are intertwined and mutually dependent.

HUMANITARIAN AGENCIES

Agencies providing outside humanitarian assistance in emergencies fall into several categories—foreign governments, international organizations, and nongovernmental organizations (NGOs) (see Annex IV).

Government Agencies

Latin American and Caribbean countries more frequently find themselves functioning as providers of humanitarian assistance than as aid recipients. In case of disasters, there often is a spontaneous show of solidarity among countries that share similar cultures and vulnerability to hazards.

Being an effective donor rather than contributing to the confusion through technically or operationally unsound initiatives should be a political priority in every country. In the Americas, the ministers of health adopted a regional policy for this purpose (see Annex III). The minimal criteria are to avoid common unsolicited donations (food, clothing, etc.) and to consult both with the affected country's ministry of health and with PAHO/WHO.

Many developed countries also offer generous bilateral assistance to disaster affected countries. Special departments or humanitarian assistance offices have been established in most donor countries. Among the most important bilateral or multilateral agencies active in the Americas are the Office of U.S. Foreign Disaster Assistance of the U.S. Agency for International Development (OFDA/USAID), which directs a comprehensive disaster mitigation, preparedness, and response program; the Office for International Humanitarian Affairs of the Canadian International Development Agency (IHA/CIDA); the United Kingdom's Department for International Development (DFID); and the European Commission Humanitarian Office (ECHO), whose budget and programs are the most significant worldwide. Other European countries and Japan have traditionally provided generous bilateral humanitarian assistance to Latin America and the Caribbean.

International Organizations

United Nations Agencies

The United Nations Office for the Coordination of Humanitarian Affairs (OCHA) is responsible for alerting the international community and coordinating international humanitarian response following all types of disasters. In addition to its coordination function, OCHA can also field a United Nations Disaster Assessment and Coordination (UNDAC) team to assist in the general assessment of needs and on-site coordination during the initial relief phase. In the Americas, a regional UNDAC team consists of qualified and specially trained nationals from PAHO Member States. This team coordinates its operations closely with the PAHO/WHO disaster team, which is activated immediately following disasters. OCHA also coordinates the dispatch of search and rescue (SAR) teams from different countries in the case of major earthquakes that affect urban areas, in order to avoid common duplications and gaps in rescue activities. Finally, OCHA coordinates the occasional multilateral deployment of military assets from a number of cooperating countries. A Military and Civil Defense Unit (MCDU) in OCHA is the focal point for civil-military cooperation, with special emphasis on the use of military assets for U.N.-led operations.

OCHA's mandate is limited to humanitarian *response*. Overall responsibility for *preparedness and mitigation* in the U.N. system has been assigned to the United Nations Development Program (UNDP), as part of the approach of integrating disaster management into the development process.

At the country level, the U.N. Disaster Management Team (DMT) is made up of representatives of all agencies of the U.N. system, including PAHO/WHO in the Americas. This team is chaired by the U.N. coordinator in the country, who is usually the UNDP Resident Representative. In some countries, the DMT also includes representatives from donor governments and NGOs. The DMT aims to offer a coordinated, multisectoral approach and collaboration to the authorities of the affected country.

In health aspects of emergencies and humanitarian assistance in the Americas, the Pan American Health Organization is the focal point and coordinator for the U.N. and inter-American systems. Its priority, however, is not to substitute local resources or to provide external material assistance but to strengthen the capacity of the countries—through preparedness and training—to respond themselves to health emergencies or disasters. In case of disaster, PAHO/WHO will provide technical cooperation in assessing health needs, formulating priorities for external health assistance, and coordinating external medical and public health external response. Although PAHO/WHO is a technical cooperation agency, it may directly provide humanitarian supplies, administer public health projects or initiatives, and offer other operational services when *no* other agency is in a position to do so. Among the technical services routinely offered is the mobilization of the expertise and capacity for management of humanitarian supplies (see Chapter 12 and Annex II for a description of SUMA, the supply management project).

Regional and Subregional Organizations in the Americas

Several subregional disaster organizations have been established during the 1990s. Decentralization from a global to a regional level is placing disaster management closer to the countries. Cooperation between neighboring countries is far preferable to the traditional international approach.

The Organization of American States (OAS) is a regional organization that lends support to its Member States in assessing their vulnerability to natural hazards and mitigating effects of disasters. FONDEM (Emergency Fund) is a mechanism established within the inter-American system for the coordination of humanitarian response among the permanent missions to the OAS, the OAS Secretariat, PAHO/WHO, the Inter-American Development Bank (IDB), and other organizations headquartered in Washington, D.C. The OAS Secretariat also offers technical assistance in risk assessment for development planning and project formulation and for reduction of vulnerability to hazards.

Following Caribbean hurricane experiences during the past few decades, and the conclusion of the Pan Caribbean Disaster Preparedness and Prevention Project in 1991, the Caribbean Governments recognized the need for a permanent regional mechanism to coordinate regional disaster management activities. The Caribbean Disaster Emergency Response Agency (CDERA) was established in 1991 by an agreement of the heads of government of the Caribbean Community (CARICOM). CDERA has 16 participating states and its headquarters are located in Barbados, West Indies. CDERA's main function is to coordinate response to any disaster affecting participating states, and to work with countries to strengthen their disaster management capacity.

In Central America, the impetus towards integration among countries resulted in the creation of the System for Central American Integration (SICA) in 1991. In their meeting in 1994, the Presidents of Central America agreed to convert the Center for Coordination for the Prevention of Natural Disasters in Central America (CEPREDENAC) into an official organization within SICA, with its headquarters in Panama. CEPREDENAC has worked since 1988 to build the capacity of scientific institutions in reducing vulnerability to disasters. The Central American Member Governments have given it the task of promoting disaster reduction in the region through an exchange of information, the development of common approaches to problem analysis, and the establishment of regional strategies to reduce disaster vulnerability.

In the Andean countries, regional cooperation regarding health issues was formalized in the Hipólito Unanue Agreement, signed in 1971. Strategies to reduce the impact of disasters on the health sector is a focus of annual meetings of the Ministers of Health of participating countries.

Nongovernmental Organizations

Worldwide, several thousand nongovernmental organizations (NGOs) are wholly or partly concerned with international humanitarian assistance, human rights, or health, and provide material, expertise, or, in some instances, cash. There are several NGO associations at the international level (see Annex IV):

- The International Council of Voluntary Agencies (ICVA), based in Geneva;
- InterAction, a Washington-based consortium of NGOs in the United States that strives to set minimum standards and promote best practices in humanitarian assistance;
- The Steering Committee for Humanitarian Response, a long-standing and influential Geneva-based working party among the International Federation for Red Cross and Red Crescent Societies, CARE International, Caritas Internationalis, Catholic Relief Services, Lutheran World Federation, Médecins Sans Frontières International, OXFAM, and the World Council of Churches;
- Voluntary Organizations in Cooperation in Emergencies (VOICE), a consortium of European agencies working in emergencies—based in Brussels, it represents European agencies before ECHO.

Many of these agencies are supported by contributions from the general public, although they increasingly receive and depend on government financing.

Nongovernmental organizations vary considerably in their approaches to humanitarian assistance and the health contributions they can make. Larger, experienced agencies and those already engaged in development work in the affected country tend to have a better understanding of the nature of the problems encountered. They engage in disaster relief only when needs have been identified. Among the most experienced agencies, national Red Cross societies and the International Federation of Red Cross and Red Crescent Societies in Geneva have been most active in disaster relief. Médecins Sans Frontières also has established a solid reputation for competent and effective public health response.

Agencies without a prior commitment to the country concerned generally have less knowledge of local problems and sometimes harbor misconceptions about the needs created by a disaster. They can thus increase the pressure on the local government by demanding operational support (for example, transportation) that would be better allocated to another agency.

In addition, "ad hoc agencies"—those formally or informally established by well intentioned but inexperienced persons in response to a particular disaster—can be a major drain on the operational resources and patience of the government of an affected country. They are generally the main source of unsolicited and unusable donations clogging the logistical chain.

Operationally, NGOs are generally more flexible and regarded as more directly responsive to people's needs than larger U.N. agencies. For this and other reasons, donor countries increasingly prefer to channel their material assistance and financial support through their national NGOs, rather than through multilateral agencies. The health services of the disaster-affected countries should recognize and adjust to this trend.

THE ARMED FORCES

National or foreign military forces are bound to play an increasing role in humanitarian assistance in Latin America and the Caribbean. National forces of the affected country—whether directly or through the civil defense system—assume a major responsibility for logistics (transport and communication) in relief operations. Aerial surveys are often made possible thanks to the local air force. The armed forces' role is essential and should be discussed and planned for before the occurrence of an emergency. Such resources should assist and not displace other traditional humanitarian players. In particular, the ministry of health should maintain its leadership and technical authority by directing what has to be done, how it should be done, and where in all health-related issues.

In a major disaster, the assistance of foreign military forces from Western countries will be offered and likely accepted. Although these forces increasingly attempt to conduct a dialogue with major humanitarian organizations, the cultural gap between the largest and most organized forces and the civilian organizations is significant and has traditionally resulted in misunderstandings and counterproductive missions. Civilian/military structures established for emergency operations tend to supplant the long-established coordination and command mechanisms that are in place in the disaster-affected countries.

The most effective way to preserve the national health authority when there is external intervention and the coordinating role of PAHO/WHO is through continuous dialogue and participation in periodic meetings or exercises organized regionally. Ignoring the issue will not improve coordination in the next disaster.

OBTAINING INTERNATIONAL DISASTER RELIEF

Most major agencies (donor governments, NGOs) have local offices to which inquiries and requests for assistance should be directed. Requests should be formulated as soon as possible after the disaster and directed to the appropriate agency. The appropriateness of an agency for meeting a specific request will depend on its resources, communication channels, and constraints.

Resources

Agencies can make cash grants, donate supplies, provide technical assistance, furnish food, or make loans. Some specialize in only one of these areas, while others have a more general mandate. It is essential to understand how each agency functions, in order to avoid requesting cash from an agency that provides only in-kind assistance, or supplies from an agency that specializes in technical cooperation. For example, PAHO/WHO, as a specialized technical agency, provides technical expertise and cooperation rather than cash or material assistance. Financial institutions (such as the Inter-American Development Bank, Caribbean Development Bank, Corporación Andina de Fomento) principally consider loans or small grants for development and reconstruction. They will, as a matter of policy, steer clear from emergency assistance of a humanitarian nature. The World Bank has established the Disaster Management Facility to provide a more strategic and rapid response to emergencies, and to promote the inclusion of risk analysis and disaster prevention mechanisms in all World Bank operations and country assistance projects.

Communication Channels

Proper communication channels are important, because some agencies may only accept requests for assistance from one specific source within the affected country, or will only disseminate assistance through a specific agency or ministry in a country. For example, PAHO/WHO accepts requests for assistance from health ministries, while the IFRCS distributes its aid exclusively through its national members, the national Red Cross Societies. However, despite these preferential channels, the ministry of health, through its Health Disaster Coordinator, should remain the ultimate public health authority in the affected country, and must be informed of and monitor the type and quantity of health assistance arriving in the country.

The ministry of health's coordinating role in managing donations has been facilitated by the adoption of SUMA, a standardized methodology for sorting, classifying, and inventorying incoming humanitarian supplies (see Chapter 12).

Decentralization of decision-making from donor agency headquarters to country offices has both facilitated the immediate on-site approval of small relief grants (generally under US$ 50,000), but has also complicated the processing of larger requests, which have to be reviewed by an increasing number of administrative levels.

Constraints

Donor agencies frequently operate within constitutional or statutory limits imposed on their activities. Some require the declaration of a state of emergency by the affected country or their own representative or a formal request from the government before they can respond. A request made to the U.N. OCHA is regarded as a request to the entire U.N. system. Most agencies must account for their programs and expenditures to a supervisory political body or public overseers, thus making projects with high visibility and humanitarian appeal (for instance, search and rescue) easier to fund than low-profile projects, such as sanitation measures.

Agencies may require first-hand or conclusive evidence of the need for relief before making expenditures or conducting fund raising, so the health ministry should arrange for agency representatives to visit disaster sites. Donors are increasingly better informed through their local experts, NGOs, or others of the validity of needs, and are less likely to blindly accept official information. For instance, blaming the natural disaster for long-standing development problems and requesting emergency humanitarian funds for their solutions is detrimental. The most valuable asset for a country and an international organization such as PAHO/WHO is its technical credibility.

Domestic public pressure will stimulate some foreign governments and agencies to commit funds or pledge support for specific projects or areas in the early stage of an emergency. This may occur prematurely, and before a thorough assessment of health sector priorities has even been initiated. The actual delivery of supplies or services may take considerably longer. The health sector must, therefore, prepare and submit preliminary cost estimates for short-term emergency humanitarian assistance needs as soon as possible before all emergency funds available are committed by donors. These estimates of immediate humanitarian needs are distinct from the estimated cost of the disaster to the health sector. These immediate humanitarian needs are, of course, much more modest in amount and must be met within the first days. Presenting to the donor community the total or conglomerate cost of the health impact (immediate needs, reconstruction cost, and indirect economic impact) is confusing, as humanitarian donors—by statute—must refrain from development or reconstruction activities.

A final constraint on some agencies is time, as their ability to respond quickly to a request for assistance varies greatly. Delays between the identification of needs by the affected country and the actual arrival of assistance from the outside are thus unavoidable and sometimes prolonged, resulting in assistance that arrives after needs have been met. Future emergency needs must therefore be anticipated in time for the external assistance to arrive: Request *today* for *tomorrow's* emergency needs (see Figure 13.1). PAHO/WHO traditionally collaborates with the countries in this task.

The decentralized nature of decision-making in donor countries also contributes to a longer approval process. Consensus should be reached between the local representative in the affected country and headquarters at home before any funding is allocated. E-mail and other electronic communication should be used more frequently to reduce delays. Access to and use of the Internet is fast becoming a necessity before and during emergency situations.

COORDINATING INTERNATIONAL HUMANITARIAN ASSISTANCE

The affected country must make clear administrative arrangements to communicate with, coordinate, and supervise the work of governmental and nongovernmental organizations. This can best be accomplished in regular meetings with representatives from all major bilateral or international agencies. The U.N. Coordinator, and in health matters, PAHO/WHO, may assist in programming these meetings and, if appropriate, offer a neutral hosting facility. In addition, donor agencies should each have a permanent liaison with the national emergency committee, thus enabling them to present problems to the committee as they arise.

As noted in Chapter 12, the government must clearly state that emergency health supplies and personnel should not be sent unless specifically requested (see Annex III). Informing donors of what is *not* wanted or needed is as critical as giving specifications for requirements. This statement should be circulated to all potential suppliers of assistance and diplomatic and consular representatives abroad. Adopting and periodically updating guidelines and procedures for diplomatic personnel abroad will prevent ineffective blanket appeals that lead to a flow of inappropriate donations.

FIGURE 13.1. Sequence of events that could potentially delay the arrival of requested supplies after a disaster.

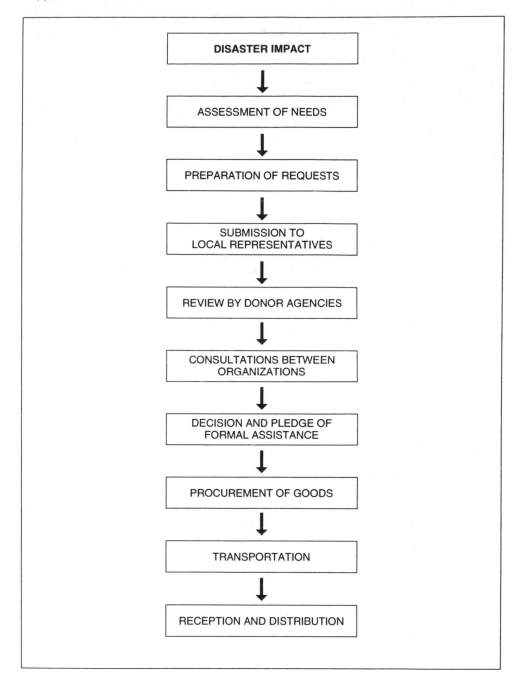

However, even when the country has clearly spelled out what is or is not required, the arrival of unsolicited medical assistance, particularly in the form of drug donations or volunteer physicians or other health personnel, may be a persistent problem. On the one hand, self-supporting teams from neighboring countries or regions that share a common culture and language can provide valuable assistance. On the other, individual foreign volunteers who are unfamiliar with local conditions, unaffiliated with a recognized agency, or in some instances, have unconfirmed academic credentials, have been most counterproductive.

The simplest but at times impractical way of dealing with this problem may be for the affected country to deny admission to any medical volunteers who arrive without institutional accreditation and support. As a corollary, if foreign medical graduates and other health workers are allowed to work in the affected country after a disaster, provisions will have to be made for temporary registration or licensing requirements to be waived and malpractice insurance, if required, will have to either be provided or waived.

Bilateral mutual assistance agreements between neighboring countries or within the framework of political integration at the subregional level offer an important opportunity for horizontal cooperation, leading to collective self-reliance in Latin America and the Caribbean. Especially in disasters affecting border areas, formal agreements and joint planning and exercises are essential to removing many of the common obstacles to a prompt and effective response.

In summary, a key to the coordination of the international response is open sharing of information. Transparency is always in the best interest of the health services and the affected country. Shortcomings and even disorganization or some degree of chaos are inherent in any disaster situation, even in the most developed country. Any attempt to hide shortcomings will only stimulate the curiosity of the international mass media and will undermine the international community's confidence in the official assessment of needs. Old models of centralized information are becoming obsolete and unsustainable in a democratic environment. Donor briefings should be organized by the health sector in consultation with civil protection and foreign affairs authorities.

Public information policy should not be restricted to providing official "statements" but should encourage NGOs and other partners to openly share their post-disaster observations through, for instance, an electronic discussion group or World Wide Web site. This exchange should not require prior "validation" or clearance by the authorities.

CHAPTER 14.
REESTABLISHING NORMAL HEALTH PROGRAMS

In the first weeks after a disaster, the pattern of health needs will change rapidly, moving from casualty treatment to more routine primary health care. Services must be reorganized and restructured, often because permanent facilities were severely damaged in terms of their infrastructure and vital systems, and severe financial constraints impede rapid reconstruction. Priorities also will shift from health care toward environmental health measures during the rehabilitation phase.

The Health Disaster Coordinator will be faced with decisions in three main areas that must not be overlooked during emergency operations: identifying and addressing long-term problems caused by the disaster; reestablishing normal health services, and assessing, repairing, and reconstructing damaged systems and buildings.

LONG-TERM HEALTH EFFECTS CAUSED BY DISASTERS

Extended Need for Medical Care

If large numbers of casualties have resulted from the disaster, a small proportion (probably less than 1%) will require long-term nursing at home, institutional care, or specialized rehabilitation for months or years. Examples are paraplegics, patients with severe brain damage, amputees, and patients with chronic sepsis. In countries where specialized services for long-term care and rehabilitation are limited, this will put a strain on the health services.

Funding long-term rehabilitation programs with international resources may prove difficult, since many organizations do not have funds for such expenditures. Preliminary statistics on the numbers of patients involved, and estimates of cost should be obtained as soon as possible and made available to both national decision makers and interested international agencies.

Surveillance of Communicable Diseases

As the weeks pass after a disaster, the public is likely to become progressively less concerned about the risk of epidemic diseases in the affected region, or the reemergence of diseases such as tuberculosis or malaria. Disease surveillance remains important and should be continued until normal disease reporting systems can be restored.

Care of Displaced Populations

A major disaster with high mortality leaves a substantial displaced population, among whom are those requiring extensive medical treatment and orphaned children. When it is not possible to locate relatives who can provide care, orphans may become the responsibility of health and social agencies. Assistance should be sought from national Red Cross societies who have considerable experience in identifying orphaned children.

Efforts should be made to reintegrate disaster survivors into society as quickly as possible through institutional programs coordinated by ministries of health and social welfare, education, and work, and family welfare institutions. Special needs of the disaster survivors should be addressed when making proposals for rehabilitation and reconstruction projects presented to international financial institutions, international agencies, and nongovernmental organizations.

REESTABLISHING NORMAL HEALTH SERVICES

Two problems may occur in restoring health services to the level existing before the disaster. In some cases, resources budgeted for six months or a year are depleted in a few days of emergency relief operations, and in disasters attracting massive voluntary or international assistance, the emergency level of services or care may temporarily exceed what the country can normally afford.

The Health Disaster Coordinator should keep in mind anticipated rehabilitation needs when formulating requests for assistance, considering requirements for the affected region before the occurrence of the disaster and the short-term needs of the affected population. In most cases, outside assistance will not exceed 10% of the total requirements, and most needs will have to be satisfied locally. Accepting certain types of assistance, such as equipment, medicines, and voluntary personnel, should also be decided in terms of long-term needs.

The rehabilitation period provides an excellent opportunity for making major changes in health care delivery, since during this phase decision-makers are receptive to new ideas. For instance, clinical laboratory services, epidemiologic surveillance, oral rehydration for diarrhea patients, expanded immunization, and maternal and child health programs have been strengthened as an indirect result of floods in the Region of the Americas.

ASSESSMENT, REPAIR, AND RECONSTRUCTION OF DAMAGED FACILITIES AND LIFELINES

When water supply and sewerage systems, hospitals, and other health facilities have been damaged, the engineering sector must arrange for an immediate survey to determine damage and functionality of the facilities, including cost estimates for repair or reconstruction of the damaged facilities and systems. Existing hazards and risks should be considered when carrying out damage assessments, so that appropriate mitigation measures can be applied when repairs and reconstruction are carried out, thus avoiding damage in future disasters.

If international assistance is required for reconstruction, project plans should be as accurate and detailed as possible, including vulnerability analysis studies, and be submitted as soon as possible, since this will increase the likelihood of obtaining funds.

Annex I.
Implementing a National Disaster Mitigation Program for Hospitals

The objectives of this strategy are to ensure the performance of hospitals in the aftermath of a disaster (in this example, an earthquake). The strategy aims to:

- Reduce vulnerability,
- Implement the plan at a reasonable cost,
- Ensure the continuity of service.

1. ASSEMBLE A TEAM OF EXPERTS

This multi-disciplinary team should include engineers (structural, mechanical, civil, and sanitation engineers), architects, seismologists, etc. The team is responsible for considering the following aspects:

- Structure,
- Architectural elements,
- Lifelines,
- Equipment,
- Organization,
- Characteristics of the area surrounding the hospital (population, access, supporting infrastructure).

2. DESCRIBE THE HEALTH SYSTEM

Analyze the overall health system in terms of its past development and current organization. The analysis should include public sector facilities (ministry of health, social security, and military hospitals) and private hospitals. Interaction among the facilities and the level of complexity should be detailed.

3. ESTIMATE THE HAZARD

In the case of seismic hazards, general and local seismicity should be determined in terms of maximum intensities and local effects. If this type of information is available for the region, it will be possible to estimate ground acceleration and expected displacement, and to establish design spectra.

The useful life of a hospital should be taken into account (30 years is a reasonable figure, both in terms of structural and functional characteristics). The level of acceptable risk is also defined (based on technical, economic, social, and political criteria).

4. CONDUCT A PRELIMINARY VULNERABILITY ANALYSIS

The first phase of the analysis is to prioritize the hospitals to be analyzed and to select the most appropriate strategy. Training should be carried out in a variety of sectors to ensure that the analysis is completed rapidly.

The vulnerability of a facility is quantified in terms of its structural, non-structural, and organizational, or functional elements.

5. SELECT BUILDINGS FOR ANALYSIS

Priority is given to highly vulnerable structures in high risk zones.

6. MAKE A QUANTITATIVE EVALUATION OF STRUCTURES

This is a detailed analysis, and solutions are recommended based on specific standards.

- Review architectural, structural, and installation diagrams;
- Compare diagrams and structure to verify whether construction actually followed the original design;
- Analyze quality and characteristics of construction materials;
- Use mathematical models depending on seismic resistant classification;
- Calculate cost-effectiveness of retrofitting.

7. PRIORITIZE THE INVESTMENT IN PROJECTS

To prioritize projects for investment, the team must consider organizational, political, technical, and financial criteria. If there are not sufficient resources to implement measures in all hospitals, the work can be programmed in phases.

8. PRODUCE A DETAILED RETROFITTING PLAN AND ARRANGE FINANCING

In this phase, the retrofitting plan for a specific project is produced, taking into account that the hospital must remain operational while construction is under way. Ideally, financing should come from national sources.

9. EXECUTE THE MITIGATION PROJECT

Annex II.
SUMA—A Humanitarian Supply Management System

The flood of relief supplies that arrive in the aftermath of a large-scale disaster often pose serious logistic and management problems for national authorities. To address these problems, the Pan American Health Organization, in conjunction with other international agencies and governments, initiated the Supply Management Project, known as SUMA, in 1992.[1] The main objective of this project is to strengthen national capacity to effectively manage humanitarian assistance supplies, from the moment donors commit to sending supplies, to the arrival and distribution of supplies at the site of a disaster. To this end, thousands of officers in more than 30 countries in the Americas and in other regions have received training.

In most countries in the Region of the Americas, SUMA focal points have been designated to coordinate the project. Among the institutions involved in the project are: ministries of health and other health agencies, civil defense or national emergency agencies, ministries of foreign relations, customs departments, Red Cross Societies, fire fighters, and nongovernmental organizations involved in humanitarian assistance.

In the immediate aftermath of large-scale disasters, especially in smaller countries, it may be unrealistic to count on local trained health professionals to sort through incoming medical supplies. PAHO/WHO provides logistical and technical support in mobilizing SUMA teams from nearby countries.

One of the most important features of SUMA is its flexibility. It can be used in many different emergency situations, and for response to natural disasters as well as in complex emergencies. The development and modification of the software has depended on constant feedback from national team members who have used it in a variety of disaster situations and training sessions.

---

[1]The SUMA software is copyrighted by PAHO, but is distributed free of charge in English, Spanish, and French. Copies of SUMA software and manuals are available on request from the Emergency Preparedness Program, PAHO/WHO, 525 23rd St., NW, Washington, DC 20037, USA; Fax (202) 775-4578; e-mail: disaster@paho.org, or from FUNDESUMA, Aptdo. 114, Plaza Mayor 1225, San José, Costa Rica, Fax: (506) 257-2139; e-mail: funsuma@sol.racsa.co.cr. The software and manuals also can be downloaded from the SUMA Web site (http://www.disaster.info.desastres.net/SUMA/) where announcements, information on emergencies, and related material can be viewed. Information on SUMA training can be obtained from the above addresses or PAHO/WHO Country Offices in countries of the Region of the Americas.

HOW DOES SUMA WORK?

SUMA team members attend a three-day course, after which they are able to apply SUMA in a disaster situation. The teams sort and label supplies, and employ user-friendly software to create an inventory of supplies and provide reports to disaster managers on the availability and distribution of items.

The system comprises three modules. The Central Level Module is set up at the Emergency Operations Center; the Field Unit Module is the basic data collection unit and operates at the points where supplies arrive during an emergency; and the Warehouse Management Module assists warehouse managers in stock control and distribution to peripheral levels. Another module assists in the management of requests to and offers from donors. Running SUMA software requires an IBM compatible 386 (or faster) computer, with 4 MB of available RAM, and 10 MB of available hard drive.

SORTING AND LABELING SUPPLIES

Information on supplies is collected at different points of entry of the disaster-affected country (e.g., airport, seaport, borders). Items are classified by category, subcategory, and item. Depending on the needs of disaster victims, supplies are sorted into different priority levels and labeled. Adhesive tapes, printed in English, Spanish, and French, are applied to each package received, showing three levels of priority. Urgently needed, or priority 1, items receive red labels, marked "Urgent! Immediate Distribution." Priority 2 supplies, that are useful but not urgently needed, receive blue labels, marked "Non-Urgent Distribution." Priority 3 items, which are of no use or require major time and effort to separate and classify, receive black labels marked "Non-Priority Items." There is space on the labels for writing the contents of the package, its weight, and destination.

After classifying supplies, their technical characteristics, potency, presentation, packaging units, total quantity, etc., are forwarded in electronic format to the central level (the Emergency Operations Center). Standard or customized reports can be easily generated for disaster coordinators, assisting them to monitor pledges from donors and identify gaps or duplications.

DISTRIBUTION AND STORAGE OF SUPPLIES

Once items have been sorted, classified, and inventoried, they are delivered to the consignees, or they are sent to an existing or temporary storage facility. SUMA teams work at warehouses and distribution hubs, managing information on the distribution of items from central to peripheral sites. All information regarding the distribution is transferred to the Emergency Operations Center.

ANNEX FIGURE II.1. Graphic overview of SUMA.

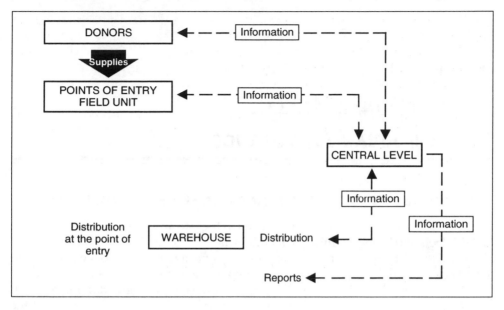

ANNEX III.
INTERNATIONAL HEALTH
HUMANITARIAN ASSISTANCE

A REGIONAL POLICY FOR INTERNATIONAL RELIEF ASSISTANCE

After the traumatic disasters experienced in Mexico and Colombia in 1985, high-level delegates of the governments of Latin America and the Caribbean met in Costa Rica in 1986 with representatives from international agencies, donor countries, and NGOs to examine ways to make international health relief assistance more compatible with the needs of affected communities. The recommendations made at this meeting—approved unanimously by the participants—form the basis of the regional policy for the Pan American Health Organization (PAHO) regarding international health relief assistance. The essence of this policy, to which all Member Governments of PAHO adhere, is the following:

- Foreign health humanitarian assistance should be made only in consultation with officials designated by the ministry of health to coordinate such assistance.
- National health authorities should quickly assess needs for external assistance and immediately alert the international community to the specific type of assistance which is, or is not, needed. Priorities should be clearly stated, distinguishing between immediate needs and those destined to rehabilitation and reconstruction.
- Diplomatic and consular missions should communicate to donor countries firm policies on the acceptance of unsolicited or inappropriate supplies.
- To avoid duplication of health humanitarian assistance, full use should be made of PAHO's clearinghouse function to inform donors of pledged contributions and determine genuine health needs.
- Countries should give high priority to the preparation of their own health and medical personnel to respond to the emergency needs of the affected population. Donor countries and organizations should support such disaster preparedness activities.
- All countries must identify their vulnerability to disasters and establish appropriate measures to mitigate the impact on the most vulnerable populations.

MAKING DISASTER ASSISTANCE EFFECTIVE

International humanitarian assistance, if provided effectively, can play an important role in a country's development. The following are suggestions to donors on how to avoid past mistakes and make international health assistance truly effective:

- **Don't stereotype the disaster.** The effects of disasters on health differ according to the type of a disaster, the economic and political situation in the affected country, and the degree to which the country's infrastructure has developed.
- **It is unlikely that medical personnel will be required from abroad,** given the capacity of Latin America and the Caribbean to mobilize health resources to respond to the immediate needs of disaster victims. In recent disasters, local health personnel treated injuries within the first 24 hours.
- **The need for search and rescue, life-saving first aid, and other immediate medical procedures is short-lived.** Special caution is necessary when considering international assistance that is useless once the acute emergency phase has passed. This type of assistance includes personnel, specialized rescue equipment, mobile hospitals, and perishable items.
- **International donors should not compete with each other to meet the most visible needs of an affected country.** The quality and appropriateness of the assistance is more important than its size, monetary value, or the speed with which it arrives.
- **Emergency assistance should complement, not duplicate, efforts taken by the affected country.** Some duplication is unavoidable as many countries and agencies worldwide hasten to meet the same needs, real or presumed. However, this need not have negative consequences if the assistance can be used later for rehabilitation and reconstruction.
- **Don't overreact to media reports for urgent, immediate international assistance.** Despite tragic images that may be shown, wait to get the overall picture and until pleas for aid have been formally issued.

ANNEX IV.
EXTERNAL AGENCIES PROVIDING HEALTH HUMANITARIAN ASSISTANCE

Every country is a potential source of health humanitarian assistance for some other disaster-stricken nation. Bilateral assistance, whether personnel, supplies, or cash, is probably the most important source of external aid. Several intergovernmental or regional agencies have established special funds, procedures, and offices to provide humanitarian assistance.

This annex uses selected examples to illustrate the broad variety of extra-national agencies that provide health assistance after natural disasters. It is not intended to be a comprehensive list, and not all experienced and dedicated agencies providing valuable emergency assistance are included. Additional information and links to other humanitarian agencies are available on the Web sites listed.

UNITED NATIONS AGENCIES

United Nations Office for the Coordination of Humanitarian Affairs (OCHA)

The United Nations plays an important role in providing assistance in response to major humanitarian emergencies, as well as in promoting disaster reduction as part of the development plans of countries. The UN Office for the Coordination of Humanitarian Affairs (OCHA), which replaced the Department of Humanitarian Affairs in 1998, coordinates the UN System's response to major humanitarian emergencies, both natural and man-made, and promotes action to improve disaster prevention and preparedness. OCHA's responsibilities after disaster are, at the request of the disaster-stricken country, to assess needs, issue inter-agency appeals for funding humanitarian assistance, organize donor meetings and follow-up arrangements, monitor the status of contributions in response to appeals, and issue reports regarding developments.

The Resident Representative of the United Nations Development Program (UNDP) in individual countries reports to OCHA, and provides a channel for requests from governments to the international community. In addition, United Nations disaster management teams, country-level representatives of the U.N. agencies have been established in many countries, make arrangements to coordinate relief activities in anticipation of an emergency.

To permit rapid response to emergencies, OCHA has established a United Nations Disaster Assessment and Coordination Team (UNDAC), which can be deployed immediately to an affected country to help local and national authorities determine relief requirements and carry out coordination.

New York office: OCHA, United Nations, S-3600, New York, NY 10017, USA
Geneva office: OCHA, United Nations, 8-14 ave. de la Paix, 1211 Geneva 10, Switzerland
Website: http://www.relifweb.int/ocha_ol

World Health Organization (WHO)

WHO is responsible for coordinating international health action. The Pan American Health Organization (PAHO) and other WHO regional offices act as focal points for national health authorities and donors after disasters in their respective areas.

WHO can provide technical cooperation in assessing health-related needs, coordinating international health assistance, managing the inventory and distribution of relief supplies (see Annex II), carrying out epidemiologic surveillance and disease control measures, assessing environmental health, managing health services, formulating cost estimates for assistance projects, and procuring humanitarian supplies. WHO and its regional offices can provide limited material assistance by reprogramming country development activities or from other sources.

WHO, Avenue Appia 20, 1211 Geneva 27, Switzerland
Website: http://www.who.int/eha

United Nations Children's Fund (UNICEF)

While primarily concerned with building health, education, and welfare services for children and mothers in developing countries, UNICEF also has mechanisms to meet their immediate needs in emergencies. Working closely with U.N. agencies and NGOs, UNICEF emergency interventions focus on the provision of health care, nutrition, water supply and sanitation, basic education, and the psychosocial rehabilitation of traumatized children. UNICEF has a substantial cash reserve for use in emergencies, allowing the diversion of funds from regular programs to emergency operations pending the receipt of donor contributions.

UNICEF, 3 United Nations Plaza, New York, NY 10017, USA
Website: http://www.unicef.org

World Food Program (WFP)

The WFP furnishes large amounts of foodstuffs in support of economic and social development projects in developing countries. In addition, it has substantial resources with which to meet emergency food needs, some of which can be furnished from project food stocks already in a disaster-stricken country. The WFP purchases and ships food needed in emergencies on behalf of donors, and cooperates closely with WHO in the nutritional monitoring of emergencies.

World Food Program, Via Cesare Giulio Viola, 68, Parco dei Medici, Rome 00148, Italy
Website: http://www.wfp.org

Food and Agriculture Organization of the United Nations (FAO)

The FAO provides technical cooperation and promotes investment in long-term agricultural development. It also works to prevent food shortages in the event of widespread crop failures or disasters. Through the Global Information and Early Warning system, the FAO issues monthly reports on the world food situation. Special alerts identify, for governments and relief organizations, countries threatened by food shortages. In both relief and short-term rehabilitation operations, FAO specialists are called on to help farmers re-establish production following floods, outbreaks of livestock disease, and similar emergencies.

FAO, Viale dell Terme di Caracalla, 1-00100 Rome, Italy
Website: http://www.fao.org

INTERGOVERNMENTAL ORGANIZATIONS

European Community Humanitarian Office (ECHO)

The European Union established ECHO in 1992 to oversee and coordinate humanitarian operations in non-member countries. ECHO works in partnership with NGOs, specialized United Nations agencies, and international bodies such as the International Committee of the Red Cross. In its first five years of existence, ECHO distributed emergency and reconstruction aid to areas of crisis in more than 60 countries. ECHO provides an important part of the operating budgets for humanitarian assistance for specialized U.N. agencies, and is the second largest donor to the World Food Program. It provides emergency aid, food aid, and aid to refugees and displaced people, in addition to investing in disaster prevention projects in high-risk regions.

ECHO, Rue Belliard 232, 1040 Brussels, Belgium
Website: http://europa.eu.int/comm/echo

Organization of American States (OAS)

The OAS is a regional agency that lends support to its Member States in assessing their vulnerability to natural hazards and mitigating the effects of disasters. It is active in technical assistance in development planning and project formulation and training projects. The OAS operates the Inter-American Fund for Assistance in Emergency Situations (FONDEM), which is administered by representatives from the OAS, the Inter-American Development Bank, and PAHO. Subject to the availability of voluntarily contributed funds, FONDEM provides food, medical supplies, and other relief to OAS Member States affected by disaster.

Center of Coordination for the Prevention of Natural Disasters in Central America (CEPREDENAC)

CEPREDENAC, an official organization within the System for Central American Integration, has worked since 1988 to build the capacity of institutions in Central America to reduce vulnerability to disasters. With headquarters in Panama, it promotes disaster reduction in the region through exchange of information, developing common approaches to problem analysis, and developing regional strategies. In the aftermath of disasters, CEPREDENAC provides technical assistance in assessment and rehabilitation efforts.

CEPREDENAC, Aptdo. Postal 3133 Balboa, Ancón, Panama
Website: http://www.cepredenac.org

Caribbean Disaster Emergency Response Agency (CDERA)

CDERA is an intergovernmental regional disaster management organization established in 1991 by the Caribbean Community (CARICOM). CDERA has 16 participating states and has its headquarters in Barbados. CDERA's main function is to coordinate response to any disaster affecting participating states. Types of assistance provided or coordinated by CDERA include relief supplies, emergency communications, emergency management personnel, and financial assistance. CDERA also works with countries to strengthen their disaster management capacity.

Caribbean Disaster Emergency Response Agency, The Garrison, St. Michael, Barbados
Website: http://www.cdera.org

NONGOVERNMENTAL ORGANIZATIONS

Adventist Development and Relief Agency (ADRA)

In 1983, the Seventh-day Adventist World Service was reorganized under the name Adventist Development and Relief Agency. Active in development projects in 143 countries, ADRA also provides humanitarian assistance in disaster situations in the form of medical assistance, shelter, emergency supplies, and technical assistance.

ADRA Central Office, 12501 Old Columbia Pike, Silver Spring, MD 20904, USA
Website: http://www.adra.org

American Council for Voluntary International Action (InterAction)

InterAction is a coalition of some 150 US-based, non-profit international development, disaster relief, and refugee assistance agencies. InterAction conducts advocacy campaigns on behalf of its members, coordinates and promotes relief and development activities, and operates as an information clearinghouse.

InterAction, 1717 Massachusetts Ave. NW, Suite 801, Washington, DC 20036, USA
Website: http://www.interaction.org

CARE (Cooperative for Assistance and Relief Everywhere)

CARE International is a confederation of 10 national members in North America, Europe, Japan, and Australia. Based in Belgium, it manages more than 340 relief and development projects in 62 countries in Africa, Asia, Latin America, and Eastern Europe. CARE USA, which oversees projects in Latin America, is based in Atlanta and provides emergency relief in the form of food, hand tools, and similar goods to disaster-affected communities. Its postdisaster projects include rehabilitation of water supply systems, rebuilding houses, and provision of basic sanitation or health facilities.

CARE USA, 151 Ellis Street, NE, Atlanta, GA 30303-2439, USA
Website: http://www.care.org

CARITAS Internationalis

CARITAS Internationalis is an international confederation of 146 Catholic organizations in 194 countries and territories. It promotes, coordinates, and supports emergency relief and long-term rehabilitation activities.

CARITAS Internationalis, Palazzo San Calisto 16, I-00120 Citta del Vaticano, Vatican
Web site: http://www.caritasint.org

Catholic Relief Services (CRS)

CRS, based in the United States, responds rapidly to emergencies by providing food, clothing, medical supplies, and shelter. Assistance is coordinated with the national CARITAS organization and the local Catholic clergy. CRS employs health professionals such as public health advisers and nutritionists who work closely with national health authorities.

Catholic Relief Services World Headquarters, 209 W. Fayette St., Baltimore, MD
    21201-3443, USA
Website: http://www.catholicrelief.org

International Committee of the Red Cross (ICRC)

ICRC is a private, Swiss, and strictly neutral humanitarian organization based in Geneva. It works to protect and assist victims of armed conflict or disturbances. If a natural disaster should befall war refugees, for example, ICRC can provide aid in kind and services, particularly nutritional and medical assistance.

ICRC, 19 Ave. de la Paix, 1202 Geneva, Switzerland
Website: http://www.icrc.org

International Council of Voluntary Agencies (ICVA)

ICVA is an international association of nongovernmental, not-for-profit organizations who are active in the fields of humanitarian assistance and development cooperation. It does not implement relief or development projects itself, but provides an international liaison structure for voluntary agency consultation and cooperation.

ICVA, 48, chemin de Grand-Montfleury, 1290, Versoix, Switzerland
Website: http://www.icva.ch

International Federation of Red Cross and Red Crescent Societies (IFRC)

IFRC is an international humanitarian organization, composed of and representing 175 member national societies, with an international secretariat based in Geneva. It coordinates humanitarian assistance internationally and operates within an affected country through the member national society or its own staff if no local society exists. The IFRC obtains cash donations and specific emergency items through international appeals, and donates them through the national society.

Assistance provided by IFRC or national societies consists of food, shelter, water and sanitation, medical supplies, telecommunications, volunteer workers, and, in some cases, self-supporting field hospitals and medical teams. Its long experience and considerable flexibility and resources make it a most valuable nongovernmental source of support and cooperation with the health sector.

IFRC, PO Box 372, CH1211 Geneva 19, Switzerland
Website: http://www.ifrc.org

Lutheran World Relief Federation (LWR)

LWR represents Lutheran churches of various denominations in the United States. It can provide in-kind assistance following disasters as well as loans for long-term reconstruction.

Lutheran World Relief, 390 Park Avenue South, New York, NY 10016, USA
Website: http://www.lwr.org

Médecins Sans Frontières (MSF)

In 1971, a group of French doctors established MSF, a humanitarian aid organization that provides emergency medical assistance to vulnerable populations in more than 80 countries. In countries where health structures are insufficient or even non-existent, MSF collaborates with national health authorities, working in rehabilitation of hospitals and pharmacies, vaccination programs, and water and sanitation projects. In addition to providing medical teams, MSF transports and distributes emergency supplies.

Médecins Sans Frontières International Office, 39, Rue de la Tourelle-1040, Brussels, Belgium
Website: http://www.msf.org

Mennonite Central Committee (MCC)

MCC is the relief and development arm of the North American Mennonite and Brethren in Christ churches. Founded in 1920, MCC has more than 700 volunteers in 50 countries involved in food relief, agriculture, health, education, and social services. MCC provides volunteer personnel for cleanup, repair, and reconstruction, as well as emergency supplies in disaster situations.

Mennonite Central Committee, PO Box 500, Akron, PA 17501-0500, USA
Website: http://www.mennonitecc.ca

Oxfam (Oxford Committee for Famine Relief)

Oxfam was founded in England to send relief supplies to refugees in Europe during World War II. Today, Oxfam International is a network of 11 humanitarian organizations based in Australia, Belgium, Canada, Hong Kong, Ireland, the Netherlands, New Zealand, Quebec, Spain, the United Kingdom, and the United States. The focus of their work is to address issues of poverty, providing financial, technical, and networking assistance to grassroots groups undertaking community development. During disasters, Oxfam provides funding and technical support for immediate and long-term assistance. It has developed considerable expertise in managing refugee camps, nutritional relief, and housing projects.

Oxfam America, 26 West Street, Boston, MA 02111, USA
Website: http://www.oxfamamerica.org
Oxfam U.K., 274 Banbury Rd., Oxford, OX2 70Z, UK
Website: http://www.oxfam.org.uk

Salvation Army

Founded in 1865 in London, the Salvation Army works in more than 100 countries to provide social, medical, educational, and other community services. In disaster situations, national affiliates provide health-care assistance and emergency supplies. It also operates an emergency radio network that assists in family tracing through a network of radio ham operators.

Salvation Army International Headquarters, 101 Queen Victoria Street, London
    EC4P 4EP, UK
Website: http://www.salvationarmy.org

Save the Children Fund/Federation

Save the Children Fund (in the United Kingdom) and Federation (in the United States) are active in more than 65 countries. Involved in long-term development projects, in disaster situations they provide food, water, shelter, and other critical supplies, and assistance in reconstruction and rehabilitation of services.

Save the Children (U.S.), 54 Wilton Road, Westport, CT 06880, USA
Website: http://www.savethechildren.org
Save the Children (U.K.), 17 Grove Lane, London, SE5 8RD, UK
Website: http://www.scfuk.org.uk

Voluntary Organizations in Cooperation in Emergencies (VOICE)

VOICE is a network of European NGOs that are active in emergency aid, rehabilitation, disaster preparedness, and conflict prevention. Created in 1992, VOICE currently has about 65 members. The main purpose of VOICE is to foster links between the NGOs and facilitate their contact with the European Union, particularly ECHO.

VOICE, 10 Square Abiorix, B-10000 Brussels, Belgium
Website: http://www.oneworld.org/voice

World Council of Churches (WCC)

The Council is a fellowship of more than 332 Protestant and Orthodox denominations in 120 countries and territories around the world, with its headquarters in Geneva. Through its member churches, it provides humanitarian assistance after disasters.

World Council of Churches, PO Box 2100, 1211 Geneva 2, Switzerland
Website: http://www.wcc-coe.org

SELECTED BIBLIOGRAPHY AND ON-LINE INFORMATION SOURCES

GENERAL EFFECTS OF DISASTERS ON HEALTH

Alexander D. The health effects of earthquakes in the mid-1990s. *Disasters* 1996;20 (3):231–247.

Armenian HK, et al. Deaths and injuries due to the earthquake in Armenia: a cohort approach. *International Journal of Epidemiology* 1997;26(4):806–813.

Basikila P, et al. Public health impact of Rwandan refugee crisis: what happened in Goma, Zaire, in July 1994? *Lancet* 1995;345(February 11):339–344.

Céspedes R, Javis D, Baxter P, Prado H. Estudio de síntomas respiratorios en escolares de las zonas aledañas al volcán Poas. (Report prepared for the Ministry of Health of Costa Rica and the University of Cambridge; 1994).

De Ville de Goyet C, et al. Earthquake in Guatemala: epidemiologic evaluation of the relief effort. *Bulletin of the Pan American Health Organization* 1976;10(2):95–109.

Glass RI, et al. Earthquake injuries related to housing in a Guatemalan village. *Science* 1977;1:638–643.

Howard MJ. Infectious disease emergencies in disasters. *Emergency Medicine Clinics of North America* 1977;21(1):39–56.

Kaneda M. Injury distributions produced by natural disasters. *Asian Med J* 1994;37(10): 557–563.

Mason J, Cavalie P. Malaria epidemic in Haiti following a hurricane. *American Journal of Tropical Medicine and Hygiene* 1965;4(4):1–10.

Noji EK. Analysis of medical needs during disasters caused by tropical cyclones: anticipated injury patterns. *Journal of Tropical Medicine and Hygiene* 1993;96:1–7.

Noji EK. *The public health consequences of disasters*. New York: Oxford University Press; 1997.

Organización Panamericana de la Salud. El impacto del huracán Mitch en Centroamérica. *Boletín Epidemiológico de la OPS* 1999;19(4).

Organización Panamericana de la Salud. Repercusiones sanitarias del fenómeno El Niño. *Boletín Epidemiológico* 1998;19 (2):9–13.

Pan American Health Organization, Emergency Preparedness and Disaster Relief Coordination Program. The earthquake in Mexico. Washington, DC: PAHO; 1985. (Disaster Reports Series No. 3).

Pan American Health Organization, Secretariat of the International Decade for Natural Disaster Reduction. *A world safe from natural disasters. The journey of Latin America and the Caribbean*. Washington, DC: PAHO; 1994.

Romero A, et al. Some epidemiologic features of disasters in Guatemala. *Disasters* 1978; 2:39–46.

Rubin CH, et al. Evaluating a fluorosis hazard after a volcanic eruption (Mount Hudson, Chile). *Archives of Environmental Health* 1994;49(5):395–401.

Sarmiento JP. El Niño Southern oscillation and communicable disease in the Americas. Washington, DC: Pan American Health Organization, Emergency Preparedness and Disaster Relief Coordination Program. In press.

Staes C, et al. Deaths due to flash floods in Puerto Rico, January 1992: implications for prevention. *International Journal of Epidemiology* 1994; 23(5):968–975.

STRUCTURING HEALTH DISASTER MANAGEMENT

Céspedes RR. Programa reducción de desastres. (Report prepared for the Ministry of Health, San José, Costa Rica, July 1996; available from the Regional Disaster Information Center, ID No. CR3.1/DES.8316).

Dyer H, Sweeney V, Poncelet JL. Health and the environment. (Report prepared for the Caribbean Program Coordinator, Barbados; available from the Regional Disaster Information Center, ID No. CR3.1/DES.9597).

García GV. Preparación del sector salud para caso de sismo. (Report prepared for the Ministry of Public Health, Cuba, 1995; available from the Regional Disaster Information Center, ID No. CR3.1/DES.6734).

Heath SE, et al. Integration of veterinarians into the official response to disasters. *Journal of the American Veterinarian Medical Association* 1997;210(February 1).

Meyer MU, et al. Health professional's role in disaster planning. *American Association of Occupational Health Nurses Journal* 1995;43(5):251–262.

Noji E, Toole M. The historical development of public health responses to disasters. *Disasters* 1997;21(4):369–379.

Poncelet JP. *Overall disaster management in the Caribbean from a health perspective,* 1996. (Available from the Regional Disaster Information Center, ID No. CR3.1/DES.8944).

Poncelet JL, De Ville de Goyet C. Disaster preparedness: institutional capacity building in the Americas. *World Health Statistical Quarterly* 1996;49(1):195–196.

Prado E, Orochena J, Rodríguez C, Casco L. Comité de desastres. (Report prepared for the Ministry of Health, Managua, Nicaragua, September 1993; available from the Regional Disaster Information Center, ID No. CR3.1/DES.7506).

DISASTER PREPAREDNESS

Céspedes R, Prado H. Preparación de la comunidad para casos de desastre. (Report prepared for the Ministry of Health, San José, Costa Rica, 1994).

Churchill RE. Effective media relations. In: Noji EK, ed. *The public health consequences of disasters.* New York: Oxford University Press; 1997.

Cohen RE, Ahearn FL. *Handbook for mental health care of disaster victims.* Baltimore: Johns Hopkins University Press; 1980.

Economic Commission for Latin America and the Caribbean. *Manual for estimating the socioeconomic effects of natural disasters.* Santiago: ECLAC; 1994.

Lewis CP, Aghababiam R. Disaster planning, part I. Overview of hospital and emergency department planning for internal and external disasters. *Disaster Medicine* 1996;14(2): 439–452.

Organización Panamericana de la Salud, Programa de Preparativos y Coordinación del Socorro en Casos de Desastres. *Manual para simulacros hospitalarios*. Washington, DC: OPS; 1995.

Pan American Health Organization, Emergency Preparedness and Disaster Relief Coordination Program. *Guidelines for assessing disaster preparedness in the health sector*. Washington, DC: PAHO; 1995.

Pan American Health Organization. *Health services organization in the event of* disaster. Wasghinton, DC: PAHO;1983. (Scientific Publication No. 443).

Reed MK. Disaster preparedness pays off. *Journal of Nursing Administration* 1998; 28(6):25–31.

Russels LA. Preparedness and hazard mitigation actions before and after two earthquakes. *Environment and Behavior* 1995; 27(6):744–770.

Savage PE. Disasters and hospital planning: a manual for doctors, nurses and administrators. Oxford: Pergamon Press; 1979.

Savage PEA. *Disasters—hospital planning*. Oxford: Pergamon Press Ltd.; 1973.

United Nations Development Programme, United Nations Disaster Relief Organization. Disasters and development: Trainer's guide for the UNDP/UNDRO disaster management training program. Madison: University of Wisconsin, Disaster Management Center; 1991. (Module prepared by RS Stephenson).

World Health Organization, Emergency preparedness and response. In: *Introduction to rapid health assessment*. Geneva: WHO; 1990.

DISASTER MITIGATION IN THE HEALTH SECTOR

Applied Technology Council. *A model methodology for assessment of seismic vulnerability and impact of disruption of water supply systems (ATC-25-1)*. Redwood City, California: Applied Technology Council; 1992.

Applied Technology Council. *Earthquake damage evaluation data for California (ATC-13)*. Redwood City, California: Applied Technology Council; 1985.

Arnold C, Reitherman R. *Building configuration and seismic design*. New York: John Wiley & Sons; 1982.

Arnold C, et al. *Seismic considerations for health care facilities*. Washington, DC: FEMA; 1987. (FEMA Report No. 150, EHRS 35).

Carby BE, Ahmad R. Vulnerability of roads and water systems to hydro-geological hazards in Jamaica. *Built Environment* 21(2/3):145–153.

Centro Panamericano de Ingeniería Sanitaria. *Estudio de caso. Terremoto del 22 de abril de 1991, Limón, Costa Rica*. CEPIS: Lima; 1996. (CEPIS Publication 96.23).

Cruz MF, Acuña R. Diseño sismo-resistente del hospital de Alajuela—un enfoque integrador. (Available from the Regional Disaster Information Center, ID No. CR3.1/DES. 6914).

Earthquake Engineering Research Institute. *Nonstructural issues of seismic design and construction*. Oakland, California: EERI; 1984. (Publication No. 84-04).

Earthquake Engineering Research Institute. *Reducing earthquake hazards: lesson learned from earthquakes*. Oakland, California: EERI; 1986. (Publication No. 86-02).

Federal Emergency Management Agency. *Instructor's guide for nonstructural earthquake mitigation for hospital and other health care facilities*. Emmitsburg, Maryland: FEMA; 1988.

Federal Emergency Management Agency. *Non-structural earthquake hazard mitigation for hospitals and other care facilities.* Emmitsburg, Maryland: FEMA; 1989. (FEMA Report No. IG 370).

Guevara LT, Jones-Parra B, Cardona OD. Método para la evaluación cualitativa de la vulnerabilidad sísmica de los aspectos no estructurales en las edificaciones médico-asistenciales en zonas urbanas de Venezuela. (Proceedings of the International Conference on Natural Disaster Management, Mérida, Venezuela, 11–14 October 1996; available from the Regional Disaster Information Center, ID No. CR3/1.DES.8919.)

McGavin GL. *Earthquake hazard reduction for life support equipment in hospitals.* Riverside, California: Ruhnau McGavin and Ruhnau Association; 1986.

McGavin GL. *Earthquake protection of essential building equipment: design, engineering, and installation.* New York: Wiley; 1981.

Naciones Unidas. Comisión Económica para América Latina y el Caribe. *Manual para la estimación de los efectos socioeconómicos de los desastres naturales,* Santiago: CEPALC; 1991.

Organización Panamericana de la Salud. Programa de Preparativos y Coordinación del Socorro en Casos de Desastres. Análisis de riesgo en el diseño de hospitales en zonas sísmicas. Washington, DC: OPS; 1989.

Organization of American States, Department of Regional Development and the Environment. *Manual for natural hazard management in planning for integrated regional development.* Washington, DC: OAS; 1993.

O'Rourke TD, McCaffrey M. Buried pipeline response to permanent earthquake ground movements. In: *Proceedings of the Eighth World Conference on Earthquake Engineering.* Vol. 7;1984:215–222.

Pan American Health Organization, Emergency Preparedness and Disaster Relief Coordination Program. *Disaster mitigation guidelines for hospitals and other health care facilities in the Caribbean.* Washington, DC: PAHO; 1992.

Pan American Health Organization, Emergency Preparedness and Disaster Relief Coordination Program. *Mitigation of disasters in health facilities: evaluation and reduction of physical and functional vulnerability.* 4 volumes. Washington, DC: PAHO; 1993.

Pan American Health Organization, Emergency Preparedness and Disaster Relief Coordination Program. *Vulnerability assessment of the drinking water supply infrastructure of Montserrat.* Barbados: PAHO; 1997.

United Nations Economic Commission for Latin America and the Caribbean. *Damage resulting from the Mexico City earthquake and its repercussions on the economy of the country.* Santiago: ECLAC Division of Program Planning and Operations; 1985.

COORDINATION OF DISASTER RESPONSE AND ASSESSMENT OF HEALTH NEEDS

Bonilla C, Céspedes R, Prado H. Instrumento de evaluación de daños y análisis de necesidades para uso en caso de desastre de instalación repentina. (Master's thesis. University of Costa Rica, San José, 1994; available from the Regional Disaster Information Center, ID No. CR3.1/DES.4643).

De Boer J. Tools for evaluating disasters: preliminary results of some hundreds of disasters. *European Journal of Emergency Medicine* 1997;4:107–110.

United Nations Development Program. Disaster Management Training Program. *Disaster Assessment.* New York: UNDP.

Vlugman A. Rapid damage and needs assessment in the sanitation and solid waste sector after a disaster. (Paper presented at the Workshop on Rapid Damage and needs Assessment in Environment Health after Disasters; available from the Regional Disaster Information Center, ID No. CR3.1/DES.5636).

MASS CASUALTY MANAGEMENT

Butman AM. *Responding to the mass casualty incident. A guide for EMS personnel.* Akron, Ohio: Emergency Training; 1982.

Canada, Ministry of National Health and Welfare. Report of the Sub-Committee on Institutional Program Guidelines. Pre-hospital emergency care services; 1985.

De Boer J, Baillie TW. *Disasters—medical organization.* Oxford: Pergamon Press; 1980.

García LM. *Disaster nursing.* Rockville, Maryland: Aspen Publications; 1985.

Hafen BQ, Karren KJ, Petersen RA. *Pre-hospital emergency care and crisis intervention workbook,* 3rd Edition. Colorado: Morzon Publishing Co.; 1989.

Noto R, Hugwenard P, Larcan A. *Médicine de catastrophe,* Paris: Editions Masson; 1987.

Pan American Health Organization, Emergency Preparedness and Disaster Relief Coordination Program. *Establishing a mass casualty management system.* Washington, DC: PAHO; 1996.

Petri RW, Dyer A, Lumpkin J. The effect of prehospital transport time on the mortality from traumatic injury. *Prehospital and Disaster Medicine* 1995;10(1):24–48.

Schultz CH, Koenig KL, Noji E. A medical disaster response to reduce immediate mortality after an earthquake. *The New England Journal of Medicine* 1996;334(7):438–444.

Spirgi E. *Disaster management: comprehensive guidelines for disaster relief.* Bern: Hans Huber; 1979.

EPIDEMIOLOGIC SURVEILLANCE AND DISEASE CONTROL

American Public Health Association. *Control of communicable diseases in man,* 16th edition. Benenson, AS, ed. Washington, DC: APHA; 1995.

Malilay J. Public health surveillance after a volcanic eruption: lessons from Cerro Negro, Nicaragua. *Bulletin of the Pan American Health Organization* 1996;30(3):218–226.

Malilay J, et al. Estimating health risks from natural hazards using risk assessment and epidemiology. *Risk Analysis* 1997;17(3):353–358.

Noji EK. The use of epidemiologic methods in disasters. In: Noji EK, ed. *The public health consequences of disasters.* New York: Oxford University Press; 1997.

Seaman J. *Epidemiology of natural disasters.* Basel: S. Karger; 1984.

Wetterhall SF, Noji EK. Surveillance and epidemiology. In: Noji EK, ed. *The public health consequences of disasters.* New York: Oxford University Press; 1997.

ENVIRONMENTAL HEALTH MANAGEMENT

Assar M. *A guide to sanitation in natural disasters*. Geneva: World Health Organization; 1971.

Farrer H. Guías para la elaboración del analisis de vulnerabilidad de sistemas de abastecimiento de agua potable y alcantarrillado sanitario. Lima: CEPIS; 1996.

Organización Panamericana de la Salud. Planificación para atender situaciones de emergencia en sistemas de agua potable y alcantarillado. Washington, DC: OPS; 1993. (Cuaderno Técnico No. 37).

Organización Panamericana de la Salud, Programa de Preparativos y Coordinación del Socorro en Casos de Desastres. *Estudio de caso: vulnerabilidad de los sistemas de agua potable frente a deslizamientos*. Washington, DC: OPS; 1998.

Organización Panamericana de la Salud, Programa de Preparativos y Coordinación del Socorro en Casos de Desastres. *Manual para la mitigación de desastres naturales en sistemas rurales de agua potable*. Washington, DC: OPS; 1998.

Organización Panamericana de la Salud. *Manual sobre preparación de los servicios de agua potable y alcantarillado para afrontar situaciones de emergencia*. Washington, DC: OPS; 1991.

O'Rourke TD, McCaffrey M. Buried pipeline response to permanent earthquake ground movements. *Proceedings of the Eighth World Conference on Earthquake Engineering*. Vol. 7; 1984:215–222.

Pan American Health Organization, Emergency Preparedness and Disaster Relief Coordination Program. *Natural disaster mitigation in drinking water and sewerage systems: guidelines for vulnerability analysis*. Washington, DC: PAHO; 1998.

Pan American Health Organization. *Environmental Health Management after Natural Disasters*. Washington, DC: PAHO; 1982. (Scientific Publication No. 430).

Pan American Health Organization, Emergency Preparedness and Disaster Relief Coordination Program. *Vulnerability assessment of the drinking water supply infrastructure of Montserrat*. Barbados: PAHO; 1997.

United Nations Disaster Relief Organization. *Disaster prevention and mitigation. Sanitation aspects*. Vol. 8. New York: UN; 1982.

FOOD AND NUTRITION

Buchanan-Smith KM. Northern Sudan in 1991: Food crisis and the international relief response. *Disasters* 1994;10:16–34.

De Ville de Goyet C, Seaman J, Geiger U. *The management of nutritional emergencies in large populations*. Geneva: World Health Organization; 1978.

Gueri M. The role of the nutrition officer in disasters. *Cajanus* 1980;13:20.

McIntosh CE. Increasing food self-sufficiency for disaster preparedness in the Commonwealth Caribbean. *Cajanus* 1985;18:84–99.

Miller DC, et al. Simplified field assessment of nutritional status in early childhood. *Bulletin of the World Health Organization* 1977;55:79–86.

Nieburg P, et al. Limitations of anthropometry during acute food shortages. *Disasters* 12:253–258.

Seaman J. Food and nutrition. *Disasters* 1981;5:180–195.

Seaman J. Principles of health care. *Disasters* 1981;3:196–204.

United Nations. *How to weigh and measure children*. New York: UN; 1988.

United Nations, Protein-Calorie Advisory Group. *A guide to food and health relief operations for disasters*. New York: UN; 1977.

World Health Organization. *Emergency preparedness and response: introduction to rapid health assessment*. Geneva: WHO; 1990.

World Health Organization. *Nutrition in times of disaster*. Report of an international conference held at World Health Organization Headquarters, Geneva, September 27–30, 1988.

PLANNING, LAYOUT, AND MANAGEMENT OF TEMPORARY SETTLEMENTS AND CAMPS

Centers for Disease Control. Famine-affected, refugee, and displaced populations: recommendations for public health issues. *MMWR Morbidity and Mortality Weekly Report* 1992;41(RR-13):1–76.

Mears C, Chowdhury S, eds. *Health care for refugees and displaced people*. Oxford: OXFAM; 1994.

Prado Z. Asentamientos humanos temporales y definitivos. (Paper presented in the Primer Seminario Nacional sobre Atención de Desastres: Memorias. Guatemala, 1984; available from the Regional Disaster Information Center, ID No. CR3.1/DES-778).

United Nations High Commissioner for Refugees. *Handbook for Emergencies*. Geneva: UNHCR; 1982.

United Nations High Commissioner for Refugees. *Water manual for refugee situations*. Geneva: UNCHR; 1992.

COMMUNICATIONS AND TRANSPORT

Ferguson EW, et al. Telemedicine for national and international disaster response. *Journal of Medical Systems* 1995;19(2):121–123.

Staffa EI. *The use of Inmarsat in disaster relief and emergency assistance operations*. (Paper presented at the International Conference on Disasters and Emergency Communications, Tampere, 1991; available from the Regional Disaster Information Center, ID No. CR3.1/DES.8762).

Stephenson R, Anderson PS. Disasters and the information technology revolution. *Disasters* 1997;21(4):305–344.

MANAGEMENT OF HUMANITARIAN RELIEF SUPPLIES

Davis J, Lambert R. *Engineering in emergencies: a practical guide for relief workers*. London: Intermediate Technology Publication Ltd.; 1995.

De Ville de Goyet C. How to make information work where it is needed. *Stop Disasters: News from the IDNDR* 1994;22:3–4.

De Ville de Goyet C. Post-disaster relief. The supply management challenge. *Disasters* 1993;17(2);169–176.

De Ville de Goyet C, Acosta E, Sabbat P, Pluut E. SUMA, a management tool for post-disaster relief supplies. *World Health Statistics Quarterly* 1996;49:189–194.

Médicins Sans Frontières-France. *Aide à l'organisation d'une mission. Situation-Intervention.* Volume II, 2nd edition. Paris: Médicins Sans Frontières-France; 1994.

Médicins Sans Frontières-Holland. *Freight and transport management. Logistic Guidelines.* Module 4.4, 2nd edition. Amsterdam: Médicins Sans Frontières-Holland; 1994.

Médicins Sans Frontières-Holland. *Warehouse and stock management. Logistic Guidelines,* 4th edition draft. Amsterdam: Médicins Sans Frontières-Holland; 1996

Refugee Policy Group. *Access to food assistance: strategies for improvement.* (Working paper; 1992).

United Nations, Department of Humanitarian Affairs. *Study on emergency stockpiles,* 2nd Edition. Geneva: UN/DHA; 1994.

United Nations Children Fund. *Assisting in emergencies, a resource handbook for UNICEF Field Staff.* Geneva: UNICEF; 1986.

United Nations High Commissioner for Refugees. *Supplies and food aid handbook.* Geneva: UNHCR; 1989.

World Food Program. *Food storage* manual, 2nd Edition. Geneva: WFP; 1983.

World Health Organization. The new emergency health kit: list of drugs and medical supplies for a population of 10,000 persons for approximately 3 months. Geneva: WHO; 1990.

INTERNATIONAL HUMANITARIAN ASSISTANCE

Benini AA. Uncertainty and information flows in humanitarian agencies. *Disasters* 1997; 21(4):335–353.

Berkmans P, et al. Inappropriate drug-donation practices in Bosnia and Herzegovina, 1992 to 1996. *The New England Journal of Medicine* 1997;337(25):1842–1845.

Burkle FM, et al. Strategic disaster preparedness and response: implications for military medicine under joint command. *Military Medicine* 1996;161(August):442–447.

Gaydos JC, Luz GA. Military participation in emergency humanitarian assistance. *Disasters* 1994;109(5):601–605.

Prado Monje H, De Ville de Goyet C. Bilateral and multilateral international cooperation: the current situation of disaster preparedness and prevention activities in Latin America and the Caribbean. (Report prepared for the Pan American Health Organization, Emergency Preparedness and Disaster Relief Coordination Program; available from the Regional Disaster Information Center, ID No. CR3.1/DES.7054).

Stockton N. Defensive development? Re-examining the role of the military in complex political emergencies. *Disasters* 1996;20(2):144–148.

Suserud BO. Acting at a disaster site: view expressed by Swedish nursing students. *Journal of Advanced Nursing* 1993;18:613–620.

ON-LINE INFORMATION SOURCES

The following documentation centers maintain on-line databases of their collections on disaster-related materials. Their web sites also provide links to other sites with disaster-related information.

Regional Disaster Information Center (CRID), San José, Costa Rica

The CRID collects and catalogues publications and papers primarily in English and Spanish on disasters, and distributes them worldwide. The database is available through the Internet and on CD-ROM.
Web site: *http://www.disaster.info.desastres/crid*

ReliefWeb, United Nations Office for the Coordination of Humanitarian Affairs

ReliefWeb compiles information on humanitarian emergencies from over 300 sources, including U.N. agencies, NGOs, governments, the academic community and the media. The site contains some 20,000 documents with information dating back to 1981.
Web site: *http://www.reliefweb.int*

Natural Hazards Information Center, Boulder, Colorado, U.S.A.

This center maintains an active Web site, and maintains a library with on-line access, and a collection of documents relating primarily to disasters and social sciences.
Web site: *http://www. colorado.edu/hazards*